T0140027

Studies in Big Data

Volume 102

Series Editor

Janusz Kacprzyk, Polish Academy of Sciences, Warsaw, Poland

The series "Studies in Big Data" (SBD) publishes new developments and advances in the various areas of Big Data- quickly and with a high quality. The intent is to cover the theory, research, development, and applications of Big Data, as embedded in the fields of engineering, computer science, physics, economics and life sciences. The books of the series refer to the analysis and understanding of large, complex, and/or distributed data sets generated from recent digital sources coming from sensors or other physical instruments as well as simulations, crowd sourcing, social networks or other internet transactions, such as emails or video click streams and other. The series contains monographs, lecture notes and edited volumes in Big Data spanning the areas of computational intelligence including neural networks, evolutionary computation, soft computing, fuzzy systems, as well as artificial intelligence, data mining, modern statistics and Operations research, as well as self-organizing systems. Of particular value to both the contributors and the readership are the short publication timeframe and the world-wide distribution, which enable both wide and rapid dissemination of research output.

The books of this series are reviewed in a single blind peer review process.

Indexed by SCOPUS, EI Compendex, SCIMAGO and zbMATH.

All books published in the series are submitted for consideration in Web of Science.

More information about this series at https://link.springer.com/bookseries/11970

Ahmed A. Abd El-Latif · Christos Volos
Editors

Cybersecurity

A New Approach Using Chaotic Systems

 Springer

Editors
Ahmed A. Abd El-Latif ⓘ
EIAS Data Science Lab, College
of Computer and Information Sciences
Prince Sultan University
Riyadh, Saudi Arabia

Department of Mathematics and Computer
Science, Faculty of Science
Menoufia University
Shibin El Kom, Egypt

Christos Volos ⓘ
Department of Physics
Aristotle University of Thessaloniki
Thessaloniki, Greece

ISSN 2197-6503　　　　　　ISSN 2197-6511　(electronic)
Studies in Big Data
ISBN 978-3-030-92168-2　　　ISBN 978-3-030-92166-8　(eBook)
https://doi.org/10.1007/978-3-030-92166-8

This Springer imprint is published by the registered company Springer Nature Switzerland AG
The registered company address is: Gewerbestrasse 11, 6330 Cham, Switzerland

Preface

Simple chaotic dynamical systems are more worth than traditional mathematical efforts in terms of complexity and performance

—Ahmed A. Abd El-Latif

Although chaotic sequences are generated by deterministic process, it has been shown that they can be considered as realizations of stochastic processes. In particular, it has been shown that, under certain assumptions, a chaotic system corresponds to an invariant measure allowing to deal with many analytical computations of statistics related to chaotic times series. Indeed, chaotic sequences have many properties similar to those of random sequences. This encourages researchers to exploit these properties in various domains of cybersecurity, such as communications, hiding information, and cryptography. Especially, in spread-spectrum communication systems, chaotic sequences have been considered as multiple-access spreading codes; in most cases where the statistics of interferences between users was computed it has been supposed that these sequences are independent.

Additionally, in the last few years there has been an increasing interest in a new classification of nonlinear dynamical systems including two kinds of attractors: self-excited attractors and hidden attractors. Self-excited attractors can be localized straightforwardly by applying a standard computational procedure, while in systems with hidden attractors we have to develop a specific computational procedure in order to identify the hidden attractors because the equilibrium points do not help in their localization. Especially systems with hidden attractors could play a vital role in the design on new advanced chaotic cryptosystems.

Furthermore, nowadays the cybersecurity, which is the practice of protecting systems, networks, and programs from digital attacks is becoming more and more prevalent. Cyberattacks are usually aimed at accessing, changing, or destroying sensitive information, extorting money from users, or interrupting normal business processes. Therefore, the need of implementing effective cybersecurity measures is particularly challenging today because there are more devices than people, while the attackers are becoming more innovative.

This book is intended to provide a relevant reference for students, researchers, engineers, and professionals working in this particular area or those interested in grasping its diverse facets and exploring the latest advances on chaotic systems for cybersecurity.

We would like to sincerely thank the authors of the contributing chapters as well as reviewers for their valuable suggestions and feedback. The editors would like to thank Dr. Thomas Ditsinger (Springer, Editorial Director, Interdisciplinary Applied Sciences), Professor Janusz Kacprzyk (Series Editor in Chief), and Ms. Rini Christy Xavier Rajasekaran (Springer Project Coordinator), for the editorial assistance and support to produce this important scientific work. Without this collective effort, this book would not have been possible to be completed.

We hope you will enjoy this book and this amazing research field of chaotic systems and their applications in cybersecurity!

Ahmed A. Abd El-Latif
Prince Sultan University
Riyadh, Saudi Arabia

Menoufia University
Shibin El Kom, Egypt

Christos Volos
Aristotle University of Thessaloniki
Thessaloniki, Greece

Acknowledgements

This work was supported by the EIAS Data Science Lab, College of Computer and Information Sciences, Prince Sultan University, Riyadh, Saudi Arabia.

Contents

Editors and Contributors

About the Editors

Dr. Ahmed A. Abd El-Latif received the B.Sc. degree with honor rank in Mathematics and Computer Science in 2005 and M.Sc. degree in Computer Science in 2010, all from Menoufia University, Egypt. He received his Ph. D. degree in Computer Science & Technology at Harbin Institute of Technology (H.I.T), Harbin, P.R. China in 2013. He is an associate professor of Computer Science at Menoufia University, Egypt and School of Information Technology and Computer Science, Nile University, Egypt. He is author and co-author of more than 140 papers, including refereed IEEE/ACM/Springer/Elsevier journals, conference papers, and book chapters. He received many awards, State Encouragement Award in Engineering Sciences 2016, Arab Republic of Egypt; the best Ph.D. student award from Harbin Institute of Technology, China 2013; Young scientific award, Menoufia University, Egypt 2014. He is a fellow at Academy of Scientific Research and Technology, Egypt. His areas of interests are multimedia content encryption, secure wireless communication, IoT, applied cryptanalysis, perceptual cryptography, secret media sharing, information hiding, biometrics, forensic analysis in digital images, and quantum information processing. Dr. Abd El-Latif has many collaborative scientific activities with international teams in different research projects. Furthermore, he has been reviewing papers for 120+ International Journals including IEEE Communications Magazine, IEEE Internet of Things journal, Information Sciences, IEEE Transactions on Network and Service Management, IEEE Transactions on Services Computing, Scientific reports Nature, Journal of Network and Computer Applications, Signal processing, Cryptologia, Journal of Network and Systems Management, Visual Communication and Image Representation , Neurocomputing, Future Generation Computer Systems, etc. Dr. Abd El-Latif is an associate editor of Journal of Cyber Security and Mobility, and IET Quantum Communication. Dr. Abd El-Latif is also leading many special issues in several SCI/EI journals.

Prof. Christos Volos received his Physics Diploma, his M.Sc. in Electronics and his Ph.D. in Chaotic Electronics in 1999, 2002, and 2008, respectively, all from the Aristotle University of Thessaloniki, Greece. He has worked from 2004 to 2010 as Scientific Associate at the Technological Educational Institutes of Thessaloniki and Serres, Greece. Also, from 2010 to 2014 he has worked as a Laboratory Teaching Staff at the Laboratory of Electronics and Telecommunications of the Greek Army Academy. He currently serves as an Assistant Professor in the Physics Department of the Aristotle University of Thessaloniki. He is also a member of the Laboratory of Nonlinear Systems, Circuits & Complexity (LaNSCom) of the Physics Department of the Aristotle University of Thessaloniki and a head of the Nonlinear Systems and Applications (NoSA) research group of the Faculty of Electrical & Electronics Engineering of the Ton Duc Thang University, Vietnam. His research interests include, among others, design and study of analog and mixed-signal electronic circuits, chaotic electronics and their applications (secure communications, cryptography, robotics), experimental chaotic synchronization and control schemes, chaotic UWB communications as well as, measurement and instrumentation systems. Dr. Volos has been author or co-author in 133 papers of international journals, 42 book chapters, 66 international conferences, and 20 national conferences. He is also a co-editor of two books, and his published work has more than 2000 non-self-citations with h-index=25. He has participated as reviewer or member of international program committees in 39 international conferences, and he is reviewer in more than 100 international journals. Also, he is member of 5 international journal's editorial board, and he has been guest editor in 19 special issues of international journals. Finally, he has participated in many national or European research scientific projects.

Contributors

Ahmed A. Abd El-Latif EIAS Data Science Lab, College of Computer and Information Sciences, Prince Sultan University, Riyadh, Saudi Arabia;
Department of Mathematics and Computer Science, Faculty of Science, Menoufia University, Shibin El Kom, Egypt

Bassem Abd-El-Atty Department of Computer Science, Faculty of Computers and Information, Luxor University, Luxor, Egypt

Akif Akgul Department of Computer Engineering, Faculty of Engineering, Hitit University, Corum, Turkey

Omer Faruk Akmese Department of Computer Engineering, Faculty of Engineering, Hitit University, Corum, Turkey

Biswabibek Bandyopadhyay Chaos and Complex Systems Research Laboratory, Department of Physics, University of Burdwan, Burdwan, India

Tanmoy Banerjee Chaos and Complex Systems Research Laboratory, Department of Physics, University of Burdwan, Burdwan, India

Akram Belazi Laboratory RISC-ENIT (LR-16-ES07), Tunis El Manar University, Tunis, Tunisia

Ali Fuat Boz Department of Electrical and Electronics Engineering, Faculty of Technology, Sakarya University of Applied Sciences, Serdivan, Sakarya, Turkey

Denis Butusov Youth Research Institute, Saint-Petersburg Electrotechnical University 'LETI', Saint Petersburg, Russia

Freddy Alejandro Chaurra-Gutierrrez INAOE, Puebla, Mexico

Murat Erhan Cimen Department of Electrical and Electronics Engineering, Faculty of Technology, Sakarya University of Applied Sciences, Serdivan, Sakarya, Turkey

Claudia Feregrino-Uribe INAOE, Puebla, Mexico

Aboozar Ghaffari Department of Electrical Engineering, Iran University of Science and Technology, Tehran, Iran

Omar Guillen-Fernandez INAOE, Puebla, Mexico

Sajad Jafari Department of Biomedical Engineering, Amirkabir University of Technology (Tehran polytechnic), Tehran, Iran; Health Technology Research Institute, Amirkabir University of Technology (Tehran polytechnic), Tehran, Iran

Ioannis Kafetzis Laboratory of Nonlinear Systems - Circuits & Complexity, Physics Department, Aristotle University of Thessaloniki, Thessaloniki, Greece

P. Kiran Department of ECE, Vidyavardhaka College of Engineering, Mysuru, Karnataka, India

Hakan Kor Department of Computer Engineering, Faculty of Engineering, Hitit University, Corum, Turkey

Mohamad Afendee Mohamed Faculty of Informatics and Computing, Universiti Sultan Zainal Abidin, Terengganu, Malaysia

Lazaros Moysis Laboratory of Nonlinear Systems - Circuits & Complexity (LaNSCom), Physics Department, Aristotle University of Thessaloniki, Thessaloniki, Greece

Fahimeh Nazarimehr Department of Biomedical Engineering, Amirkabir University of Technology (Tehran polytechnic), Tehran, Iran

Erivelton G. Nepomuceno Control and Modelling Group (GCOM), Department of Electrical Engineering, Federal University of São João del-Rei, São João del-Rei, MG, Brazil

Felipe Orihuela-Espina INAOE, Puebla, Mexico

Muhammed Ali Pala Department of Electrical and Electronics Engineering, Faculty of Technology, Sakarya University of Applied Sciences, Serdivan, Sakarya, Turkey

H. T. Panduranga Department of ECE, Government Polytechnic, Turvekere, Tumkur, Karnataka, India

Gustavo Rodriguez-Gomez INAOE, Puebla, Mexico

Aceng Sambas Department of Mechanical Engineering, Universitas Muhammadiyah Tasikmalaya, Tasikmalaya, Indonesia

Esteban Tlelo-Cuautle Department of Electronics, Instituto Nacional de Astrofísica, Óptica Y Electrónica (INAOE), Tonantzintla, Puebla, México

Aleksandra Tutueva Youth Research Institute, Saint-Petersburg Electrotechnical University 'LETI', Saint Petersburg, Russia

Sundarapandian Vaidyanathan Research and Development Centre, Vel Tech University, Avadi, Chennai, India

Christos Volos Laboratory of Nonlinear Systems - Circuits & Complexity (LaNSCom), Physics Department, Aristotle University of Thessaloniki, Thessaloniki, Greece

J. Yashwanth Department of ECE, Vidyavardhaka College of Engineering, Mysuru, Karnataka, India

Jessica Zaqueros-Martinez INAOE, Puebla, Mexico

Sen Zhang School of Physics and Opotoelectric Engineering, Xiangtan University, Hunan, Xiangtan, China

A Novel Approach for Robust S-Box Construction Using a 5-D Chaotic Map and Its Application to Image Cryptosystem

Bassem Abd-El-Atty, Akram Belazi, and Ahmed A. Abd El-Latif

Abstract Data privacy and security play a vital task in our daily life, which chaotic systems are commonly utilized for designing modern cryptosystems. This chapter presents a new chaos-based substitution box (S-box) and its application for securing images. The construction of the presented S-box approach is based on a 5-D chaotic system, which the experimental results prove that the suggested S-box approach has high nonlinearity and good cryptographic characteristics. In addition, we utilize the effectiveness of the presented S-box approach in designing a novel image cryptosystem, which the experimental outcomes prove the effectiveness of the presented image cryptosystem.

Keywords Chaos-based image cryptosystem · Chaos-based S-box · Chaotic systems · Chaos-based cryptography

1 Introduction

The importance of data security and privacy is increasing with the growing digital transformation of our daily lives. In the past, users shared their secret information on an offline platform where the likelihood of data security risk was minimal, but with technological advancements in the digital age, users share their secret information on

B. Abd-El-Atty
Department of Computer Science, Faculty of Computers and Information, Luxor University, Luxor 85957, Egypt
e-mail: bassem.abdelatty@fci.luxor.edu.eg

A. Belazi
Laboratory RISC-ENIT (LR-16-ES07), Tunis El Manar University, 1002 Tunis, Tunisia

A. A. Abd El-Latif (✉)
EIAS Data Science Lab, College of Computer and Information Sciences, Prince Sultan University, Riyadh 11586, Saudi Arabia
e-mail: aabdellatif@psu.edu.sa

Department of Mathematics and Computer Science, Faculty of Science, Menoufia University, Shibin El Kom 32511, Egypt

© The Author(s), under exclusive license to Springer Nature Switzerland AG 2022
A. A. Abd El-Latif and C. Volos (eds.), *Cybersecurity*, Studies in Big Data 102,
https://doi.org/10.1007/978-3-030-92166-8_1

online platforms to perform certain activities. Digital images are commonly used to represent data, which its security can be achieved via performing image encryption or image data hiding mechanisms [1–3].

Chaotic systems or simply chaotic maps play an important role in designing modern cryptosystems. Chaotic maps can be utilized for designing S-boxes, pseudorandom number generators, image encryption, and so on [4–7]. Chaotic maps can be categorized into two major classes: One-dimensional and multi-dimensional chaotic maps. One-dimensional chaotic systems are exposed to several attacks due to their short key space and their chaotic discontinuous ranges [8].

The construction of S-boxes, for secure cryptographic applications, attracts a great deal of attention from most scholars of cryptography. S-box plays a vital factor in designing most block ciphers and a well-designed S-box ensures the nonlinearity and confusion characteristics [9]. Using the effectiveness of the 5-D hyperchaotic map presented in [10], we proposed a new chaos-based S-box scheme. The experimental results prove that the suggested S-box approach has high nonlinearity and good cryptographic characteristics.

In addition, this chapter presents a new image encryption approach using the proposed S-box scheme and the 5-D hyperchaotic map presented in [10]. In the proposed image cryptosystem, the generated chaotic sequences form the chaotic map devoted to substitute the plain image, then a specific data about the substituted image are acquired to update the initial condition of the hyperchaotic system then iterate the chaotic system again, utilize the new chaotic sequences to permutated the substituted image and construct an S-box and utilize the constructed S-box to substitute the permutated image and get the final cipher image. The experimental results prove that the effectiveness of the presented image cryptosystem.

The layout of this chapter is set as: The presented S-box approach and its performance are provided in Sect. 2, and the suggested image cryptosystem is introduced in Sect. 3, while Sect. 4 provides the experimental results of the proposed image cryptosystem.

2 Proposed Chaos-Based S-Box Mechanism

In this part, we introduced a new approach for generating S-boxes based on a 5-D hyperchaotic map [10] besides giving its performance analysis.

2.1 The Proposed S-Box

The suggested S-box method utilizes the generated chaotic sequences from the hyperchaotic map to construct the S-box. The hyperchaotic system is represented as in Eq. (1) [10].

$$\begin{cases} x_{t+1} = (-ax_t + y_t z_t - dw_t - qv_t) \bmod 1 \\ y_{t+1} = (-ay_t + (z_t - b)x_t - dw_t - qv_t) \bmod 1 \\ z_{t+1} = (1 - x_t y_t) \bmod 1 \\ w_{t+1} = (cx_t) \bmod 1 \\ v_{t+1} = (px_t) \bmod 1 \end{cases} \tag{1}$$

where x_0, y_0, z_0, w_0, and v_0 are the primary conditions, while a, b, c, d, p, and q are control parameters of the hyperchaotic system. The specific procedures of the suggested S-box mechanism are presented below.

Step 1: Choice a primary values for the initial conditions (x_0, y_0, z_0, w_0, and v_0) and control parameters (a, b, c, d, p, and q) to operate the hyperchaotic system (1) for 256 times, where the outcomes of running hyperchaotic system are five sequences (X, Y, Z, W, and V).

Step 2: Transform the chaotic sequence V into integers of range [1, 4].

$$R = \left(\text{floor}(V \times 10^6) \bmod 4\right) + 1 \tag{2}$$

Step 3: Generate sequence Q utilizing X, Y, Z, and W sequences as explained below.

$$Q_t = \begin{cases} X_t \text{ if } R_t = 1 \\ Y_t \text{ If } R_t = 2 \\ Z_t \text{ if } R_t = 3 \\ W_t \text{ if } R_t = 4 \end{cases}, \text{ for } t = 1, 2, \ldots, 256 \tag{3}$$

Step 4: Arrange the elements of Q from the smallest to the largest as vector QA and obtain the index per number of QA in Q as the generated S-box.

2.2 Performance Analysis

Generally, robust S-boxes need to satisfy six characteristics [11, 12] which are bijective, strict avalanche criterion, nonlinearity, linear approximation probability, outputs bit independence criterion, and input/output XOR distribution. In the below subsections, we provide details of these analyses to explain the efficiency of the suggested S-box. The primary parameters utilized for operating the hyperchaotic system are set as: $x_0 = 0.6416$, $y_0 = 0.8852$, $z_0 = 0.3875$, $w_0 = 0.6384$, $v_0 = 0.2519$, $a = 2$, $b = 3$, $c = 4$, $d = 7$, $p = 8$, and $q = 6$. The generated S-box using the primary key parameters is provided in Table 1.

Table 1 An S-box constructed by the suggested method

126	34	27	189	201	72	13	208	59	191	30	20	115	110	113	214
53	221	138	75	192	132	88	168	47	9	210	212	252	105	0	71
49	24	52	103	61	241	173	50	128	19	145	116	97	124	87	162
219	242	120	203	101	37	14	193	65	234	155	25	206	190	222	246
92	235	149	23	121	174	22	186	122	118	133	225	136	244	129	12
184	111	93	248	135	95	160	236	102	78	194	180	109	243	77	56
237	169	1	117	32	157	96	33	86	98	176	209	171	36	143	205
134	62	249	6	179	185	137	223	183	247	76	57	131	164	68	17
90	227	31	46	21	233	112	240	140	232	82	197	45	207	159	8
142	153	178	172	231	228	220	5	125	73	119	81	226	54	146	152
147	58	42	28	200	250	51	40	64	2	151	170	188	10	67	130
253	154	44	156	60	16	48	195	3	215	202	239	104	255	217	177
181	43	4	114	123	199	251	100	204	11	182	213	218	94	55	108
127	254	196	229	39	89	187	38	107	35	91	84	66	163	165	141
26	245	70	41	29	85	69	167	7	230	79	63	99	74	175	150
18	139	161	224	158	106	211	166	198	238	80	83	148	216	144	15

2.2.1 Bijective Criterion

For an $n \times n$ S-box, the bijective property is satisfied if all elements of the generated S-box are different from each other and in the range $[0, 2^n - 1]$. The stated S-box in Table 1 satisfies the bijective property, in which the stated S-box has 256 elements distinct from each other and in the range $[0, 255]$.

2.2.2 Nonlinearity Property

The nonlinearity property of the generated S-box and S-boxes reported in [13–18] are given in Table 2, in which the suggested S-box method has good nonlinearity property.

2.2.3 Strict Avalanche Property

When a single input bit is complemented all of the production bits change with a probability of a half is known as strict avalanche criterion (SAC). The dependence matrix of the constructed S-box is displayed in Table 3, in which the outcomes range from 0.4063 to 0.6406 and the average value is 0.5059, which is near to the optimal value 0.5. A comparison of SAC for our S-box and the further S-boxes is displayed in Table 4, in which the results demonstrate that our constructed S-box has good SAC properties.

Table 2 Nonlinearity of the presented S-box approach and other chaos-based S-boxes

S-box	Nonlinearity		
	Max	Min	Average
Proposed	108	102	105.75
Belazi and Abd El-Latif [13]	110	100	105.50
Belazi et al. [14]	108	102	105.25
Cassal-Quiroga and Campos-Cantón [15]	104	96	101.75
Cassal-Quiroga and Campos-Cantón [15]	108	96	102.25
Khan and Asghar [16]	106	98	102.00
Liu et al. [17]	108	104	105.75
Özkaynak et al. [18]	108	100	104.75

Table 3 SAC of the constructed S-box

0.4688	0.4531	0.5625	0.4375	0.5313	0.4531	0.5156	0.4219
0.5313	0.4531	0.5469	0.5000	0.5000	0.4531	0.5469	0.5469
0.5156	0.5156	0.4688	0.5156	0.5000	0.5313	0.5000	0.5469
0.5000	0.4063	0.5156	0.4688	0.4688	0.4531	0.5625	0.6406
0.5313	0.4688	0.5000	0.5469	0.5938	0.5313	0.4688	0.5938
0.5313	0.4531	0.5625	0.5156	0.4531	0.4531	0.5625	0.4844
0.5000	0.4688	0.5313	0.5625	0.5625	0.5469	0.5156	0.5781
0.5156	0.4219	0.5625	0.4219	0.4688	0.5000	0.4688	0.4688

Table 4 SAC outcomes for the presented S-box approach and other S-boxes

S-box	SAC		
	Max	Min	Avg
Proposed	0.6406	0.4063	0.5059
Belazi and Abd El-Latif [13]	0.5625	0.4375	0.5000
Belazi et al. [14]	0.5313	0.4297	0.4956
Cassal-Quiroga and Campos-Cantón [15]	0.5781	0.3906	0.5012
Cassal-Quiroga and Campos-Cantón [15]	0.6094	0.4219	0.5059
Khan and Asghar [16]	0.6719	0.3906	0.5178
Liu et al. [17]	0.5938	0.4219	0.4976
Özkaynak et al. [18]	0.5781	0.4218	0.4982

2.2.4 Output Bits Independence Property

The output bits independence criterion (BIC) indicates that all the avalanche variables should be pair-wise independent for a provided set of avalanche vectors produced by complementing a single plaintext bit. The outcomes of the presented S-box are displayed in Tables 5 and 6, in which the average values of BIC-SAC and BIC-nonlinearity are 0.5060 and 105.0714, respectively. A comparison of average values for BIC-SAC, and BIC-nonlinearity is given in Table 7, in which our S-box fulfills the BIC criterion.

Table 5 BIC-nonlinearity outcomes for the constructed S-box

–	108	106	108	102	106	106	102
108	–	104	102	104	104	100	108
106	104	–	106	108	104	106	106
108	102	106	–	106	106	104	108
102	104	108	106	–	106	102	104
106	104	104	106	106	–	106	106
106	100	106	104	102	106	–	104
102	108	106	108	104	106	104	–

Table 6 BIC-SAC values for the constructed S-box

–	0.5087	0.5176	0.5154	0.5064	0.5087	0.5176	0.5199
0.5266	–	0.5199	0.5064	0.5087	0.5176	0.5244	0.5266
0.5020	0.4728	–	0.4773	0.4840	0.5132	0.4773	0.5176
0.4997	0.4773	0.5401	–	0.5042	0.5064	0.5266	0.4975
0.5244	0.5266	0.5042	0.4952	–	0.4975	0.5132	0.4885
0.5020	0.4930	0.4952	0.4997	0.4908	–	0.4997	0.5199
0.5176	0.5154	0.4885	0.5199	0.5132	0.4818	–	0.5176
0.5109	0.5042	0.4773	0.5244	0.5244	0.4975	0.4728	–

Table 7 BIC-SAC and BIC-nonlinearity average values for the presented S-box approach and other S-boxes

S-box	BIC-SAC	BIC-nonlinearity
Proposed S-box	0.5060	105.1
Belazi and Abd El-Latif [13]	0.4970	103.8
Belazi et al. [14]	0.4996	103.8
Cassal-Quiroga and Campos-Cantón [15]	0.5066	103.4
Cassal-Quiroga and Campos-Cantón [15]	0.5050	103.5
Khan and Asghar [16]	0.4999	102.9
Liu et al. [17]	0.5032	104.5
Özkaynak et al. [18]	0.4942	103.1

Table 8 DP table for the constructed S-box

6	6	6	6	6	6	8	6	6	8	6	6	6	6	8	8
8	8	8	4	6	8	6	6	8	6	6	6	6	8	6	6
6	6	6	6	6	6	6	6	8	8	6	6	8	8	6	6
6	6	6	6	6	6	8	6	8	8	6	8	6	8	6	6
6	6	8	8	8	6	6	6	6	6	8	6	6	8	6	6
8	6	8	6	4	8	8	6	6	6	8	6	8	6	6	10
10	6	6	6	8	6	8	8	8	6	6	6	8	6	6	6
6	6	6	6	6	8	8	6	6	8	6	4	6	6	8	8
6	4	6	8	8	6	8	6	8	6	6	6	6	6	6	6
8	8	6	8	6	6	8	6	6	8	8	8	8	8	8	6
8	6	6	8	6	4	8	8	6	6	6	6	8	6	6	6
6	6	6	6	6	8	8	8	8	8	6	8	6	8	8	8
8	6	6	6	8	8	6	6	6	8	8	6	6	6	6	8
8	6	6	6	6	6	8	6	8	6	6	8	6	6	8	10
6	8	8	6	8	8	6	6	6	6	6	10	6	6	6	8
6	6	6	8	6	6	6	8	6	8	8	6	6	6	6	–

Table 9 Maximum DP values for the presented S-box approach and other S-boxes

S-box	Max DP
Proposed	0.0313
Belazi and Abd El-Latif Belazi et al. [13]	0.0468
[14]	0.0391
Cassal-Quiroga and Campos-Cantón [15]	0.0391
Cassal-Quiroga and Campos-Cantón [15]	0.0469
Khan and Asghar [16]	0.0313

2.2.5   The Input/Output XOR Distribution

The differential approximation probability (DP) of our S-box is provided in Table 8, in which the max value is 10 that means the suggested S-box method is robust alongside differential attacks. A comparison of the max DP is given in Table 9, in which the outcomes mean that the constructed S-box fulfills the DP property.

2.2.6   Linear Approximation Probability (LP)

LP is the maximum value of the imbalance of an event. A comparison of LP for our presented S-box and other chaos-based S-boxes are provided in Table 10, in which the results proved that the constructed S-box has fulfilled the LP property.

S-box	LP
Proposed	0.1484
Belazi and Abd El-Latif [13]	0.1250
Belazi et al. [14]	0.1562
Khan and Asghar [16]	0.1250

Table 10 LP outcomes for the presented S-box approach and other S-boxes

3 Proposed Chaos-Based Image Cryptosystem

Chaotic maps act an important role in modern cryptographic procedures [9]. Using the benefits of the 5-D hyperchaotic map presented in [10], we presented a novel image encryption approach for securing images. In the proposed image cryptosystem, the generated chaotic sequences from the chaotic map devoted to substitute the plain image, then specific data about the substituted image are acquired to update the initial condition of the hyperchaotic system then iterate the chaotic system again, utilize the new chaotic sequences to permutated the substituted image and construct an S-box, and utilize the constructed S-box to substitute the permutated image and get the final cipher image. The encryption processes of the proposed cryptosystem are outlined in Fig. 1, and the specific steps are given in what follows:

Step 1: Choice a primary values for the initial conditions (x_0, y_0, z_0, w_0, and v_0) and control parameters (a, b, c, d, p, and q) to operate the hyperchaotic system (1) for $h \times w$ times, where $h \times w$ is the dimension of the pristine image (P) and the outcomes of running the hyperchaotic system are five sequences (X, Y, Z, W, and V).

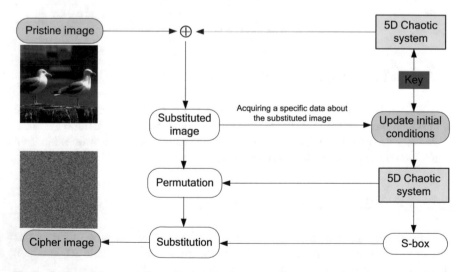

Fig. 1 Outline of the encryption procedure for the proposed image cryptosystem

Step 2:   Transform the chaotic sequence V into integers of range $[1, 4]$.

$$R = \left(\text{floor}\left(V \times 10^6\right) \bmod 4\right) + 1 \tag{4}$$

Step 3:   Generate sequence Q utilizing X, Y, Z, and W sequences as explained below.

$$Q_t = \begin{cases} X_t \text{ if } R_t = 1 \\ Y_t \text{ if } R_t = 2 \\ Z_t \text{ if } R_t = 3 \\ W_t \text{ if } R_t = 4 \end{cases}, \text{ for } t = 1, 2, \ldots, h \times w \tag{5}$$

Step 4:   Transform the generated sequence Q into integers of range $[0, 255]$.

$$K = \text{floor}\left(Q \times 10^{12}\right) \bmod 256 \tag{6}$$

Step 5:   Reshape the elements of pristine image (P) to a one vector then perform bitwise XOR operation to the reshaped image ($PVec$) and the generated key (K).

$$PVec = reshape(P, h \times w, 1) \tag{7}$$

$$SIm = PVec \oplus K \tag{8}$$

Step 6:   Acquire some information about SIm image then update the initial conditions (x_0, y_0, z_0, w_0, and v_0) of the hyperchaotic system.

$$\lambda = \frac{\left(\sum_{t=1}^{h \times w} SIm(t)\right) \bmod 1024}{2048} \tag{9}$$

$$\begin{aligned} x_{new} &= 0.5x_0 + \lambda \\ y_{new} &= 0.5y_0 + \lambda \\ z_{new} &= 0.5z_0 + \lambda \\ w_{new} &= 0.5w_0 + \lambda \\ v_{new} &= 0.5v_0 + \lambda \end{aligned} \tag{10}$$

Step 7:   Using the updated initial conditions (x_{new}, y_{new}, z_{new}, w_{new}, and v_{new}) and the primary control parameters (a, b, c, d, p, and q) operate the hyperchaotic system (1) for $h \times w$ times, where the outcomes of running the hyperchaotic system are five sequences (Xn, Yn, Zn, Wn, and Vn).

Step 8:   Transform the chaotic sequence Vn into integers of range $[1, 4]$.

$$Rn = \left(\text{floor}\left(Vn \times 10^6\right) \bmod 4\right) + 1 \tag{11}$$

Step 9: Generate sequence Qn utilizing Xn, Yn, Zn, and Wn sequences as explained below.

$$Qn_t = \begin{cases} Xn_t \, \text{if} \, Rn_t = 1 \\ Yn_t \, \text{if} \, Rn_t = 2 \\ Zn_t \, \text{if} \, Rn_t = 3 \\ Wn_t \, \text{if} \, Rn_t = 4 \end{cases}, \text{ for } t = 1, 2, \ldots, h \times w \qquad (12)$$

Step 10: Arrange the elements of Qn from the smallest to the largest as vector QA and obtain the index per number of QA in Qn as a permutation vector ($PrVc$).

Step 11: Permutate SIm image using $PrVc$ as explained below.

$$PrImV(t) = SIm(PrVc(t)), \text{for} t = 1, 2, \cdots, h \times w \qquad (13)$$

Step 12: Utilizing the first 256 elements of Qn, construct an S-box using the proposed method then substitute the permutated image ($PrImV$) via the constructed S-box for constructing the final cipher image (CIm) as explained below.

$$CImV(t) = SBx(PrImV(t) + 1), \text{for} t = 1, 2, \cdots, h \times w \qquad (14)$$

$$CIm = reshape(CImV, h, w) \qquad (15)$$

4 Performance Analyses of the Presented Image Cryptosystem

To value the suggested image cryptosystem, experimental executed on a PC with Intel[R] Core™ 2Duo, 4 GB RAM, and pre-installed with MATLAB R2016b version. The investigated dataset of images is taken from Computer Vision Group (CVG) dataset [19] and it is involve four grey scale images per of dimension 512×512 and labeled as P1, P2, P3 and P4 (see Fig. 2). The primary parameters utilized for operating the hyperchaotic system are set as: $x_0 = 0.6416$, $y_0 = 0.8852$, $z_0 = 0.3875$, $w_0 = 0.6384$, $v_0 = 0.2519$, $a = 2$, $b = 3$, $c = 4$, $d = 7$, $p = 8$, and $q = 6$.

4.1 Correlation Analysis

The general tool for measuring the meaningful of an image is the correlation coefficient of neighboring pixels [20], which its outcomes for pristine images are close to 1 ineach direction while in a cipher images of a well-designed cryptosystem should

Fig. 2 Pristine images and their corresponding ciphered versions using the proposed image cryptosystem

close to 0. To calculate the correlation coefficients of our cryptosystem, we picked randomly 10,000 pairs of neighboring pixels in every. The outcomes of the correlation coefficient are listed in Table 11 for the studied dataset, in which the outcomes for cipher images are very near to 0, while the distribution of correlation coefficient for image P1 is provided in Fig. 3. It is obvious from the outcomes stated in Table 11, and the distribution presented in Fig. 3 that no profitable data was acquired concerning the pristine image by analyzing correlations for cipher images.

Table 11 Values of correlation coefficient for the investigated images

Image	Direction		
	Horizontal	Vertical	Diagonal
P1	0.9786	0.9635	0.9458
Cipher (P1)	0.0003	−0.0001	0.0002
P2	0.9733	0.9760	0.9594
Cipher (P2)	−0.0009	−0.0002	−0.0006
P3	0.9695	0.9798	0.9602
Cipher (P3)	0.0001	−0.0003	−0.0005
P4	0.9535	0.9475	0.9149
Cipher (P4)	−0.0001	−0.0009	−0.0002

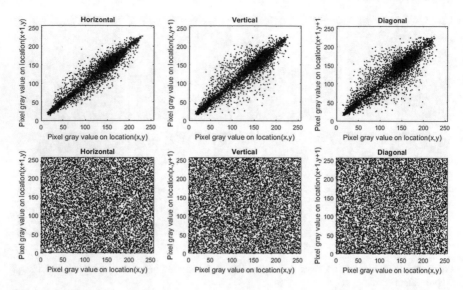

Fig. 3 Distribution of correlation for pristine and cipher P1 image, in which the distribution for the pristine image is given in top row while the second row provides the distribution for the cipher version

Table 12 Results of NPCR test for the investigated images

Image	NPCR (%)
P1	99.6227
P2	99.6048
P3	99.6159
P4	99.6021

4.2 Pristine Image Sensitivity

To value the results of slight changes of pixels in the pristine image on its equivalent cipher version, the NPCR ("Number of Pixels Change Rate") tool [21] is applied. The results of NPCR for the investigated images are listed in Table 12 when changing only the least significant bit of the pixel located in location (1, 1) for every given pristine image. The results stated in Table 12 evident that the presented encryption method has a high sensitivity to slight pixel modifications in the pristine image.

4.3 Histogram Analysis

To estimate the distribution of the organization of pixels in an image, Histogram test is performed, in which a well-designed encryption approach must have an identical

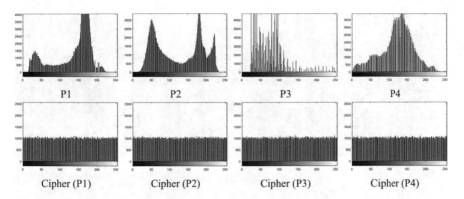

<table>
P1 | P2 | P3 | P4
</table>

Cipher (P1)   Cipher (P2)   Cipher (P3)   Cipher (P4)

Fig. 4 Histograms of the investigated images

Table 13 Outcomes of information entropy

Image	Pristine	Cipher
P1	7.12375	7.99928
P2	7.45696	7.99934
P3	5.84836	7.99929
P4	7.05810	7.99927

distribution for different cipher images to withstand statistical at-tacks. Figure 4 plots the histograms of the investigated images, in which the histograms for pristine images are different from each other while the distributions of their corresponding cipher version are indistinguishable from each other.

4.4 Entropy Analysis

To assess the concentration of the pixel values of each plane in an image, information entropy is executed [21], which the optimal entropy value is equivalent to 8-bit. Table 13 stated the results of entropy values for pristine images and their corresponding cipher images, which the results for the cipher images are very near to 8.

4.5 Occlusion Analysis

When data are transmitted over communication channels, they are easily affected by occlusion attacks [22]. Consequently, it is critical functionality for a well-designed image cryptosystem to have the aptitude to withstand occlusion attacks. To appraise

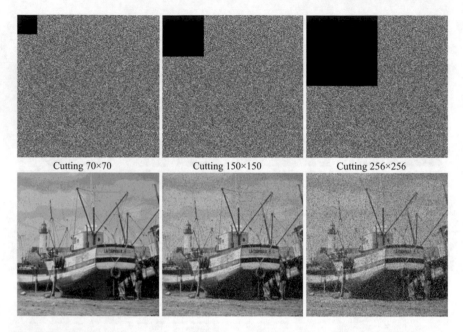

Fig. 5 Results of occlusion attack, in which the top row displays the affected cipher images with cutting data blocks, while the bottom row displays the decrypted images of the corresponding affected cipher ones

our cryptosystem against data loss attacks, we produced a cutting block with different dimensions of pixels to the cipher image P1 and then attempt to decrypt it. Figure 5 displays the results of these attempts, in which the decrypted images have good visual quality without losing visual data at the location of the cutting part.

4.6   Key Sensitivity

To evidence the key sensitivity of the displayed cryptosystem, we performed the decryption process on the ciphered P1 with slight changes in the initial values, in which the results are displayed in Fig. 6.

5   Conclusion

In this chapter, we proposed a new chaos-based S-box and presented its application for securing images. The experimental outcomes of the presented S-box approach proved its efficiency and have high nonlinearity and good cryptographic properties.

The specific keys

The specific keys except for $x_0=0.6416000000000001$

The specific keys except for $y_0= 0.885200000000001$

The specific keys except for $z_0= 0.387500000000001$

The specific keys except for $w_0= 0.6384000000000001$

The specific keys except for $v_0= 0.2519000000000001$

The specific keys except for $a= 2.000000000000001$

The specific keys except for $b= 3.000000000000001$

The specific keys except for $c= 4.000000000000001$

The specific keys except for $d= 7.000000000000001$

The specific keys except for $p= 8.000000000000001$

The specific keys except for $q= 6.000000000000001$

Fig. 6 Effects of decryption process for the cipher image P1 with tiny changes in the initial values

Also, the experimental outcomes of the proposed image cryptosystem proved its effectiveness and can be reliable in various cryptographic applications.

Acknowledgements This work was supported by Luxor University and Menoufia University, Egypt.

References

1. Abd EL-Latif, A.A., Abd-El-Atty, B., Abou-Nassar, E.M., Venegas-Andraca, S.E.: Controlled alternate quantum walks based privacy preserving healthcare images in internet of things. Optics Laser Technol. **124**, 105942 (2020)
2. Hamza, R., Yan, Z., Muhammad, K., Bellavista, P., Titouna, F.: A privacy-preserving cryptosystem for IoT E-healthcare. Inf. Sci. **527**, 493–510 (2020)
3. Kaissis, G.A., Makowski, M.R., Rückert, D., Braren, R.F.: Secure, privacy-preserving and federated machine learning in medical imaging. Nat. Mach. Intell. **2**(6), 305–311 (2020)
4. Abd El-Latif, A.A., Abd-el-Atty, B., Amin, M., Iliyasu, A.M.: Quantum-inspired cascaded discrete-time quantum walks with induced chaotic dynamics and cryptographic applications. Sci. Rep. **10**(1), 1–16 (2020)
5. Özkaynak, F.: Construction of robust substitution boxes based on chaotic systems. Neural Comput. Appl. **31**(8), 3317–3326 (2019)
6. Wang, X., Zhang, W., Guo, W., Zhang, J.: Secure chaotic system with application to chaotic ciphers. Inf. Sci. **221**, 555–570 (2013)
7. Li, L., Abd-El-Atty, B., Elseuofi, S., Abd El-Rahiem, B., Abd El-Latif, A.A.: Quaternion and multiple chaotic systems based pseudo-random number generator. In: 2019 2nd International Conference on Computer Applications and Information Security (ICCAIS), pp. 1–5. IEEE (2019)
8. Abd-El-Atty, B., Amin, M., Abd-El-Latif, A., Ugail, H., Mehmood, I.: An efficient cryptosystem based on the logistic-chebyshev map. In: 2019 13th International Conference on Software, Knowledge, Information Management and Applications (SKIMA), pp. 1–6. IEEE (2019)
9. El-Latif, A.A.A., Abd-El-Atty, B., Belazi, A., Iliyasu, A.M.: Efficient chaos-based substitution-box and its application to image encryption. Electronics **10**(12), 1392 (2021)
10. Vaidyanathan, S., Sambas, A., Abd-El-Atty, B., Abd El-Latif, A.A., Tlelo-Cuautle, E., Guillén-Fernández, O., et al.: (2021). A 5-D multi-stable hyperchaotic two-disk dynamo system with no equilibrium point: circuit design, FPGA realization and applications to TRNGs and image encryption. IEEE Access **9**:81352–81369
11. Hussain, I., Gondal, M.A.: An extended image encryption using chaotic coupled map and S-box transformation. Nonlinear Dyn. **76**(2), 1355–1363 (2014)
12. Tang, G., Liao, X., Chen, Y.: A novel method for designing S-boxes based on chaotic maps. Chaos Solitons Fractals **23**(2), 413–419 (2005)
13. Belazi, A., Abd El-Latif, A.A.: A simple yet efficient S-box method based on chaotic sine map. Optik **130**, 1438–1444 (2017)
14. Belazi, A., Khan, M., Abd El-Latif, A.A., Belghith, S.: Efficient cryptosystem approaches: S-boxes and permutation-substitution-based encryption. Nonlinear Dyn. **87**(1), 337–361 (2017)
15. Cassal-Quiroga, B.B., Campos-Cantón, E.: Generation of dynamical S-boxes for block ciphers via extended logistic map. Math. Probl. Eng. (2020)
16. Khan, M., Asghar, Z.: A novel construction of substitution box for image encryption applications with Gingerbreadman chaotic map and S 8 permutation. Neural Comput. Appl. **29**(4), 993–999 (2018)

17. Liu, G., Yang, W., Liu, W., Dai, Y.: Designing S-boxes based on 3-D four-wing autonomous chaotic system. Nonlinear Dyn. **82**(4), 1867–1877 (2015)
18. Özkaynak, F., Çelik, V., Özer, A.B.: A new S-box construction method based on the fractional-order chaotic Chen system. SIViP **4**(11), 659–664 (2016)
19. Dataset of standard 512 × 512 grayscale test images.: https://ccia.ugr.es/cvg/CG/base.htm. Accessed 12 Jul 2021
20. Abd-El-Atty, B., Iliyasu, A.M., Alanezi, A., Abd El-latif, A.A.: Optical image encryption based on quantum walks. Opt. Lasers Eng. 138, 106403 (2021)
21. Alanezi, A., Abd-El-Atty, B., Kolivand, H., El-Latif, A., Ahmed, A., El-Rahiem, A., et al.: Securing digital images through simple permutation-substitution mechanism in cloud-based smart city environment. Secur. Commun. Netw. (2021)
22. Tsafack, N., Iliyasu, A.M., De Dieu, N.J., Zeric, N.T., Kengne, J., Abd-El-Atty, B., et al.:. A memristive RLC oscillator dynamics applied to image encryption. J Inf Secur Appl **61**, 102944 (2021)

An Image Compression-Encryption Algorithm Based on Compressed Sensing and Chaotic Oscillator

Aboozar Ghaffari, Fahimeh Nazarimehr, Sajad Jafari, and Esteban Tlelo-Cuautle

Abstract In this chapter, a chaotic oscillator is presented. Various dynamical behaviors of the oscillator are analyzed. The complex dynamics of the proposed oscillator are applied in a compression-encryption algorithm. Here, we propose an image compression-encryption method using compressed sensing and a chaotic oscillator. At first, the original image is represented in the wavelet domain to obtain sparse coefficients. The sparse representation is scrambled with a chaotic sequence. The scrambling operation increases the security level and improves the performance of sparse recovery in the decryption process. The sparse scrambled representation is then compressed using the chaotic dynamics. The compressed matrix is also scrambled to reduce the elements' correlation. To obtain an unrecognizable encrypted image, the XOR operation is used. In the decryption process, the smoothed l_0 norm (SL0) algorithm decreases the complexity of calculations for image reconstruction. Wiener filter is used in sparse recovery based on SL0 to improve image reconstruction. The results of the presented method are satisfying in various compression ratios. Security analysis illustrates the effectiveness of our method.

Keywords Image compression-encryption method · Chaotic dynamics · Dynamical properties · Sparse representation · Wiener filter · Smoothed ℓ_0 norm · Measurement matrix

A. Ghaffari (✉)
Department of Electrical Engineering, Iran University of Science and Technology, Tehran, Iran
e-mail: aboozar_ghaffari@iust.ac.ir

F. Nazarimehr · S. Jafari
Department of Biomedical Engineering, Amirkabir University of Technology (Tehran polytechnic), Tehran, Iran

S. Jafari
Health Technology Research Institute, Amirkabir University of Technology (Tehran polytechnic), Tehran, Iran

E. Tlelo-Cuautle
Department of Electronics, Instituto Nacional de Astrofísica, Óptica Y Electrónica (INAOE), Tonantzintla, Puebla 72840, México

1   Introduction

Chaotic dynamics are very complex, making them worthy for various applications like image encryption and secure communication [1–3]. There are many obscurities on how the chaotic dynamics are generated in the flows [4–6]. Previously, the relation between chaotic dynamics and saddle equilibrium points was a hot topic. However, many chaotic oscillators without saddle equilibria have been proposed till now [7, 8]. Proposing new chaotic flows with various features has been a hot topic since it can help researchers learn more about chaotic dynamics [9–13]. Memristive oscillators are a subject with increased research interest [14–16]. A memristive chaotic flow was studied in [17]. Bifurcation and FPGA of a memristor oscillator were discussed in [18]. In [19], the multi-stability of a memristive oscillator was investigated. Control of a chaotic oscillator was discussed in [20]. Chaotic analog and digital circuits are very interesting [21–23]. Fractional-order chaotic systems were studied in [24–26].

Recently, images are used in various applications. The networks and the internet are necessary tools to transmit the data. These data channels are usually insecure because of the various attacks by hackers; the data should be protected in applications such as health care [27–29]. Image encryption converts data to a meaningless form to protect data from different attacks [30, 31]. So, a keystream is used to encrypt images. The encryption methods are classified into symmetric and asymmetric [31–33]. To encrypt images, different structures have been proposed such as XOR operation [34, 35], quantum image encryption [36, 37], DNA coding [38–40], and optical image encryption [41–43]. Chaotic dynamics are usually an important component of these approaches to generate pseudorandom sequences. The initial conditions of chaotic oscillators have the role of keystream [44]. Various methods can study the histogram flatness, such as chi-square and entropy [45–48]. The plaintext sensitivity is a common test for the effectiveness of an encryption method [49].

The image encryption methods can be categorized into two groups. First, some methods convert the original image to a meaningless or noisy one. Second, an encrypted image is visually meaningful, while the original image is hidden and trans-ferred by a carrier image [50–52]. Recently, with the development of compressed sensing, another category of encryption has been presented. The compression and encryption are combined in this category due to limited bandwidth in the trans-mission network [48, 53–57]. This class has considered the sparse representation of the original image in the transform domain. The sparse representation has been compressed by a measurement matrix. The matrix was usually generated by a chaotic oscillator. The original images were recovered using an optimization approach of sparse recovery. Usually, to increase the security level, the traditional approach and compressed sensing approach are combined. Encryption based on chaotic map [58–68], discrete fractional random transform [69], cellular automata [59, 70], and visually meaningful encryption schemes [51, 52] are some examples.

An important measure to evaluate compressed sensing-based encryption approaches is reconstruction performance. Three effective parts of this category were transform, measurement matrix, and sparse recovery method. In the first step of the approach, the original image was transformed to another domain such as wavelet [63, 65] to obtain sparse representation. These transforms can be block-based [70] or global-based [65, 71]. The scrambling of sparse coefficients before compression via measurement matrix was used [60]. To scramble the sparse coefficients, researchers have proposed different approaches such as the zigzag scan [58–60] and random perturbation based on chaotic sequence [48]. The compression step based on the measurement matrix was also categorized into 1D and 2D approaches. In the 1D approaches, a measurement matrix is applied to compress the sparse matrix from one direction. 2D methods use two measurement matrices based on the sparse decomposition of 2D time series [72], or Kronecker compressed sensing [73, 74]. This class has reduced the complexity of calculations and has required memory to recover the original image [72–74]. Measurement matrix can be constructed via different approaches [58, 66–68] and orthogonal random matrix based on chaotic oscillators [48]. Usually, these matrices, which are keystream in image encryption, are generated by chaotic oscillators.

Here, a chaotic oscillator is proposed. Then an image encryption approach based on compressed sensing and the chaotic oscillator is presented. This approach contains six steps, transformation of the original image, chaotic confusion of sparse coefficients, compression via measurement matrix, quantization and mapping, and the final encryption using chaotic confusion and XOR operation. In the proposed approach, a random permutation using the proposed chaotic oscillator is used to increase the image reconstruction performance. The final encryption step based on scrambling and XOR is used to obtain a meaningless image with minimum correlation among adjacent pixels. In this chapter, the performance of two compression approaches (1D and 2D) are compared. In the decryption process, the smoothed ℓ_0 norm (SL0) is used [72, 75], which is a fast and accurate method to recover the original image. Since the quantization step adds some noise to the measured and compressed encrypted image, this chapter investigates the effect of this noise via the robustness analysis of sparse reconstruction. Here, to decrease this noise's effect and improve the SL0 algorithm, the Wiener filter is combined with the SL0. In fact, in the proposed sparse recovery approach, it is assumed that the recovered image has smooth variations; hence this information is applied to the sparse recovery optimization via the Wiener filter. Experimental results show that this idea improves compression and reconstruction performance. Security analyses investigate the results of the image encryption. Image encryption has been a hot topic recently. The limited bandwidth in the insecure transmission network is a crucial challenge in this area. So, we propose a combination of encryption and compression algorithms to deal with this challenge.

In Sect. 2, the sparse recovery is described. Section 3 illustrates the chaotic oscillator. The encryption method is stated in Sect. 4. Section 5 provides the experimental results and security analyses. In Sect. 6, the chapter is concluded.

2 Sparse Representation

Signal decomposition (SD) is a basic approach to extract the important information of the signal structure. SD provides a representation of a vector signal $y \in R^n$ via a linear combination of some basis signals d_i, $1 \le i \le m$ as follows:

$$y = s_1 d_1 + \cdots + s_m d_m = Ds \tag{1}$$

where $D = [d_1, \ldots, d_m] \in R^{n \times m}$, and $s = [s_1, \ldots, s_m]^T \in R^m$. Traditional transforms such as wavelet and DCT are special complete ($m = n$) SD cases of the above decomposition. When the matrix D is underdetermined $n < m$, the representation s is not unique. Among the feasible set of this linear combination, the sparsity assumption (the smallest number of non-zero elements) on the representation s is used frequently in many applications such as compressed sensing [76, 77], pattern recognition [78, 79], image registration [80, 81], and image denoising [82, 83]. The goal of sparse recovery in these applications is finding the sparsest solution of this linear combination as follows:

$$P_0^{1D} : \text{Minimize} \|s\|_0 \ s \cdot t \ y = Ds \tag{2}$$

where $\|s\|_0$ stands for the ℓ_0 norm of s. It has been shown that if the linear Eq. (1) is sparse enough, then this is a unique solution. The following theorem states the uniqueness condition using spark.

Theorem (Uniqueness [84]): Let $spark(D)$ show the minimum number of linearly dependent columns of D. If the equation $y = Ds$ has a solution s_0 for which $\|s_0\| < spark(D)/2$, it is its unique sparsest solution.

A critical challenge in the sparse recovery is finding the sparsest solution. This optimization is generally NP-hard because the ℓ_0 norm is not differentiable and sensitive to noise. Researchers have proposed several methods to solve this problem, for example, SL0 [75].

Images as 2D data are used frequently in many applications. In traditional approaches, to apply compressed sensing for data, the 2D problem is transformed into a 1D problem. This approach has two challenges of the high required memory and computational complexity. To resolve this problem, researchers have proposed an approach using two measurement matrices, $D_1 \in R^{n_1 \times m_1}$ and $D_2 \in R^{n_2 \times m_2}$, into two directions to compress a sparse matrix $S \in R^{m_1 \times m_2}$ in the following mathematical form.

$$P_0^{2D} : \text{Minimize} \|S\|_0 \ s \cdot t \ Y = D_1 S D_2^T \tag{3}$$

This problem is a particular case of 1D sparse recovery (2) [72]. It can be shown that the sparsest solution of this optimization is unique under some conditions related to the spark values of two measurement matrices [72, 74, 85]. Some practical solvers have been proposed to find the sparsest solution of 2D compressed sensing [72, 86, 87].

In this chapter, the SL0 algorithm is used as a fast and accurate algorithm in the decryption process. SL0 solves the ℓ_0 norm directly by replacing ℓ_0 norm with a smooth approximation of ℓ_0, like:

$$\|s\|_0 \approx \|s\|_{\sigma 0} = n - \sum_{i=1}^{n} \exp\left(-\frac{s_i^2}{2\sigma^2}\right)$$

where n is the number of elements. The smoothed ℓ_0 norm is equivalent to ℓ_0 norm in $\sigma \rightarrow 0$. This smoothed sparsity measure is optimized iteratively by decreasing σ [72, 75]. In this chapter, to improve the reconstruction performance of the SL0 approach, the smoothness assumption of the original image is added to the sparse recovery problem using the Wiener filter. Results show that our method outperforms the other ones.

3 The Proposed Chaotic Oscillator

Here a novel chaotic oscillator is proposed as Eq. (4).

$$
\begin{aligned}
x &= z \\
\dot{y} &= z - y \\
\dot{z} &= \eta y + z + 0.3x^2 - 1.5xy + 0.6xz
\end{aligned}
\tag{4}
$$

The oscillator shows chaotic solutions in parameter $\eta = -1$ and initial values $[-40, -2.85, -3.22]$. Figure 1 shows the 3D chaotic dynamic and its projections. To investigate the system dynamics, in the first step, its equilibrium points should be calculated. The oscillator has an equilibrium point in $[0, 0, 0]$. The oscillator has zero eigenvalues. So the eigenvalues cannot characterize the stability of the equilibrium point. Numerical studies show the unstable form of the equilibrium point.

The bifurcation diagram of the oscillator concerning changing η and using the backward continuation method is shown in Fig. 2. It is plotted using the Poincare section of the z variable (the peak values of the z variable). The figure shows that in the interval $\eta \in [-5, -3.6]$, the system goes to an invariant torus around the origin. Then in $\eta = -3.6$, the dynamics jump to a period-1 limit cycle and continue in the interval $[-3.6, -2.475]$. In $\eta \in [-2.475, -1.59]$, the oscillator presents a period-three limit cycle and continues with a period-doubling route to chaos. The Lyapunov spectrum of the oscillator by backward continuation initiation is plotted in Fig. 3. Part (a) of the figure shows three Lyapunov exponents, and part (b) presents two larger Lyapunov exponents.

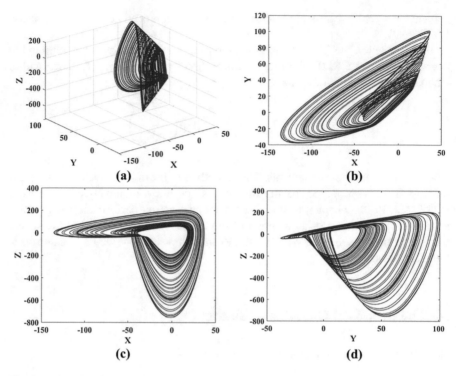

Fig. 1 a 3D chaotic dynamics of the oscillator in parameter $\eta = -1$ and initial values $(-40, -2.85, -3.22)$ and the projections in **b** $X - Y$; **c** $X - Z$; **d** $Y - Z$

Fig. 2 Bifurcation diagram of the oscillator by varying η and backward continuation method

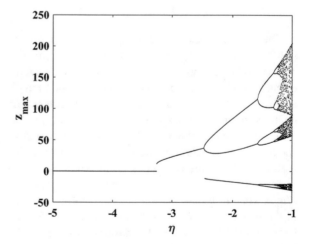

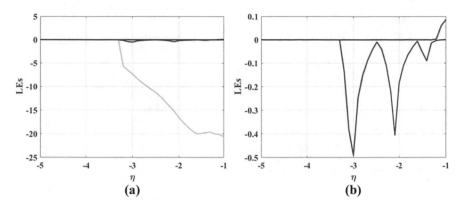

Fig. 3 Lyapunov exponents (LEs) of the oscillator by changing η, **a** three LEs; **b** two larger LEs

Various properties of the oscillator show the complexity of its dynamics. So, the oscillator is a suitable choice for the application of image encryption.

4 The Proposed Encryption Method

Here, a new approach of image compression-encryption is presented. This approach contains three main blocks as follows:

1. In the first block, the original image is represented in the wavelet transform to obtain sparse representation. Then the sparse coefficients are shuffled to increase the security level and the performance of sparse recovery in the decryption part.
2. In the second block, the sparse representation is compressed via the measurement matrices and then quantized and mapped in the interval [0, 255].
3. In the third block, the compressed image is encrypted via pixel shuffling and XOR operation using the chaotic sequence. In this step, a meaningless image with minimum correlation among adjacent pixels is obtained.

Figure 4 shows the block diagram of the image compression-encryption algorithm. Before starting the proposed approach, two blocks of scrambling samples and constructing a measurement matrix using the chaotic oscillator are demonstrated. In summary, the wavelet transform is applied to the original image. The obtained representation is scrambled and compressed using the measurement matrix. The compressed image is mapped in the interval [0, 255]. Finally, the mapped image is encrypted again using the XOR and scrambling. In other words, the original image is sequentially encrypted by two steps of compressed sensing and XOR operation.

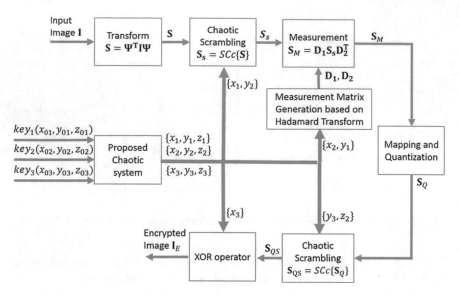

Fig. 4 The block diagram of the proposed encryption approach

4.1 Chaotic Scrambling

Scrambling the entries of a matrix is used in two steps of the presented approach to increase the security level by decreasing the correlation between matrices' elements. Here, to scramble the entries of matrix $\mathbf{M} \in R^{n \times m}$, the rows and columns are shuffled separately via the random sequences of the proposed chaotic oscillator. The chaotic scrambling (CSc) steps to shuffle the entries of matrix $\mathbf{M}$ are as follows.

1. Generating the sequences $X = \{x_i\}$ and $Y = \{y_i\}$ of the proposed chaotic oscillator: The lengths of X and Y are equal to the number of elements of the matrix $\mathbf{M}$.
2. Generating two random integer sequences by converting the sequences $X = \{x_i\}$ and $Y = \{y_i\}$ into integer sequences $X^* = \{x_i^*\}$ and $Y^* = \{y_i^*\}$:

$$x_i^* = \mathrm{mod}(x_i \times 10^{12}, 2^{16})$$

$$y_i^* = \mathrm{mod}(y_i \times 10^{12}, 2^{16})$$

3. Sorting the sequences X^* and Y^*:

$$\left[Z_s^*, idx_s\right] = \mathrm{sort}\left(X^*Y^*\right)$$

where $[.,.] = \text{sort}(.)$ is the sequencing operator, Z_s^* is the sorted sequence. idx_s is the index of Z_s^*. In fact, idx_s is an integer sequence with a random perturbation obtained from sorting the random integer sequence X^*Y^*.

4. Shuffling the matrix $\mathbf{M}$ elements with the sequence idx_s.

4.2 Chaotic Measurement Matrix Generation Based on Hadamard Transform

The measurement matrix is the basic of compressed sensing. This matrix compresses and measures the original signal in low dimension via the linear combination. Here, Hadamard transform and chaotic oscillator construct the measurement matrix. This matrix is a part of the key that increases the security level. In this chapter, two measurement matrices $\mathbf{D}_1 \in R^{n_c \times n}$, and $\mathbf{D}_2 \in R^{m_c \times m}$ compress the sparse representation $\mathbf{S} \in R^{n \times m}$ in the form $\mathbf{S}_M = \mathbf{D}_1\mathbf{S}\mathbf{D}_2^T \in R^{n_c \times m_c}$. Compression ratios in the directions of row and column are equal to $\frac{n_c}{n}$ and $\frac{m_c}{m}$, respectively. Hence, the total compression ratio (CR) is equal to $\frac{n_c m_c}{nm}$. $\mathbf{D}_1$ and $\mathbf{D}_2$ matrices are generated with the same approach, except they use different chaotic sequences, as mentioned in Fig. 4. The generation process of matrix $\mathbf{D}_1 \in R^{n_c \times n}$ is described as follows:

1. Generating the sequences $Y = \{y_i\}$ of the proposed chaotic oscillator: Y's lengths are equal to n_c.
2. Obtaining a random integer sequence $Y^* = \{y_i^*\}$:

$$y_i^* = \text{mod}(y_i \times 10^{16}, 2^{16})$$

3. Sorting the sequence Y^*:

$$[Y_s^*, idx_s^Y] = \text{sort}(Y^*)$$

idx_s^Y is a sequence of integer numbers with random perturbation.
4. Generating the Hadamard matrix $\mathbf{H} \in R^{n \times n}$.
5. Shuffling the rows of the matrix $\mathbf{H}$ with the sequences idx_s^Y and obtaining the matrix $\mathbf{H}_s$.
6. Calculating the measurement matrix $\mathbf{D}_1$ with the first n_c rows of the matrix $\mathbf{H}_s$, i.e. $\mathbf{D}_1 = \mathbf{H}_s(1 : n_c, :)$.

4.3 Image Compression-Encryption Method

Here, the encryption process (Fig. 2) is described as follows:

Step 1: The representation of the image $I \in R^{n \times m}$ in the transform domain $\mathbf{\Psi}$ such as wavelet is computed to calculate the sparse representation S:

$$\mathbf{S} = \mathbf{\Psi}^{\mathrm{T}}\mathbf{I}\mathbf{\Psi} \in R^{n \times m}$$

Step 2: The entries of the sparse matrix $\mathbf{S}$ are shuffled by the operator of the chaotic scrambling ($CSc\{.\}$):

$$\mathbf{S}_s = CSc\{\mathbf{S}\}$$

Step 3: Two measurement matrices compress the scrambled sparse matrix S_s

The sparse matrix S_s is calculated using two chaotic matrices $\mathbf{D}_1 \in R^{n_c \times n}$ and $\mathbf{D}_2 \in R^{m_c \times m}$:

$$\mathbf{S}_M = \mathbf{D}_1\mathbf{S}_s\mathbf{D}_2^T \in R^{n_c \times m_c}$$

In this chapter, two cases of compression are compared. In the first case, it is assumed that the CR in the column direction is equal to ones, i.e., $m_c = m$. On the other hand, the total CR is equal to the CR in the row direction. This case is called Compression-Encryption based on 1D compressed sensing (CE-1DCS). In the second case, the compression ratios in two directions are identical, i.e., $\frac{m_c}{m} = \frac{n_c}{n}$. This case is called Compression-Encryption based on 2D compressed sensing (CE-2DCS). The experimental results present that CE-2DCS has better performance than CE-1DCS.

Step 4: Mapping and quantizing the measured matrix in the interval [0, 255] to represent each entry of the compressed matrix by one byte to prepare this for the final encryption based on XOR:

$$\mathbf{S}_Q = \left[\frac{255(\mathbf{S}_M - \min(\mathbf{S}_M))}{\max(\mathbf{S}_M) - \min(\mathbf{S}_M)} \right]$$

where $[x]$ is the round of the entry of x to the nearest integer value.

Step 5: Reducing the correlation among adjacent pixels by scrambling the quantized matrix $\mathbf{S}_Q$:

$$\mathbf{S}_{QS} = CSs\{\mathbf{S}_Q\}$$

This process is performed using the chaotic scrambling algorithm.

Step 6: The scrambled and compressed sparse representation $\mathbf{S}_{QS}$ is XORed with a random integer sequence by the proposed chaotic sequence by the following steps:

1. Generating the sequence $X = \{x_i\}$ of the proposed chaotic oscillator: Y's lengths are equal to $n_c m_c$.
2. Converting the sequence Y to a random integer sequence $X^* = \{x_i^*\}$:

$$x_i^* = \mod(x_i \times 10^{10}, 2^8)$$

3. In the final step, the matrix $\mathbf{S}_{QS}$ is encrypted by the XOR operator to have a meaningless image with a uniform histogram:

$$\mathbf{I}_E = \mathbf{S}_{QS} \oplus X^*$$

To do this operation, the matrix and random sequence must be converted in the binary domain, and after the XOR operation, the encrypted image is converted to the decimal format.

In the presented approach, the proposed chaotic oscillator is run with two different initial conditions.

4.4 The Decryption Process

In this section, the decryption method is described. This process inverses all steps of the proposed encryption to obtain the original image. It is assumed that the keystream containing initial values of the chaotic oscillator is available to calculate required sequences. Since the proposed encryption process contains two sequential steps of encryption, the inverse of these steps are investigated as follows,

- **Decryption 1**: This step inverses the XOR operation and chaotic scrambling:

1. The encrypted image I_E is XORed to obtain the compressed and scrambled compressed representation $\mathbf{S}_{QS}$:

$$\mathbf{S}_{QS} = \mathbf{I}_E \oplus Y^*$$

2. The mapped compressed matrix $\mathbf{S}_Q$ is computed via the inverse chaotic scrambling operator $SCs^{-1}\{.\}$:

$$\mathbf{S}_Q = SCs^{-1}\{\mathbf{S}_{QS}\}$$

- **Decryption 2**: In this step, the compression process is inverted via the sparse recovery. Three inverse steps of this decryption are investigated as follows:

1. Inverse mapping to obtain the measured matrix $\mathbf{S}_M$:

$$\mathbf{S}_M = \frac{\mathbf{S}_Q(\max(\mathbf{S}_M) - \min(\mathbf{S}_M))}{255} + \min(\mathbf{S}_M)$$

Note that $\max(\mathbf{S}_M)$ and $\min(\mathbf{S}_M)$ are also sent.
2. In this step, the scrambled sparse matrix $\mathbf{S}_s$ is computed via the sparse recovery problem (2) or (3) for the cases of compression CE-1DCS or CE-2DCS. Note that in this application, CE-1DCS is a particular case of

CE-2DCS. In this chapter, SL0 is used to recover sparse representation. Here, the performance of the sparse reconstruction is improved by adding the Wiener filter to the SL0 algorithm. In the proposed decryption method, sparse recovery, Wiener filter, and decryption of scrambling are combined. The following section describes the proposed decryption.

3. The sparse representation $\mathbf{S}$ is obtained via the inverse chaotic scrambling operator:

$$\mathbf{S} = SCs^{-1}\{\mathbf{S}_s\}$$

4. At the final step, the image is recovered by the inverse transform $\mathbf{\Psi}$,. i.e.

$$\mathbf{I} = \mathbf{\Psi}\mathbf{S}\mathbf{\Psi}^{\mathbf{T}}$$

For the decryption process, the key parameters, including initial conditions of the chaotic attractors, compression ratio, $\min(S_M)$, and $\max(S_M)$, should be sent to the receiver.

4.5 Sparse Recovery Based on Smoothed ℓ_0 Norm and Wiener Filter

In this section, a sparse recovery approach based on smoothed ℓ_0 norm and Wiener filter is proposed. Since CE-1DCS is a particular case of CE-2DCS, the optimization of problem CE-2DCS is described in this application. The cost function of SL0 is a smoothed approximation of the ℓ_0 norm as follows:

$$\underset{\mathbf{S}_s}{\text{argmin}} J_\sigma(\mathbf{S}_s) = \sum_{i,j}\left(1 - \exp(-\frac{s_{ij}^2}{2\sigma^2})\right), s.t \mathbf{S}_M = \mathbf{D}_1\mathbf{S}_s\mathbf{D}_2^T \qquad (5)$$

The cost function $J_\sigma(\mathbf{S}_s)$ is equivalent to ℓ_0 norm when $\sigma \rightarrow 0$. In [72, 75], the approach of graduated nonconvexity (GNC) [88] is used to obtain the sparse solution. This approach uses a decreasing sequence of σ to optimize the cost function $J_\sigma(\mathbf{S}_s)$ iteratively. In this method, the cost function $J_\sigma(\mathbf{S}_s)$ is minimized for a fix σ, and its solution is applied as a start point to minimize $J_\sigma(\mathbf{S}_s)$ for the smaller σ. This method helps the SL0 algorithm to escape from local minima. In SL0, the problem (5) for a fixed σ is optimized by the steepest descent approach. Each of its iterations $(\mathbf{S}_s \leftarrow \mathbf{S}_s - \mu \nabla_{\mathbf{S}_s} J_\sigma(\mathbf{S}_s))$ is followed by projection onto the feasible set $\{\mathbf{S}_M | \mathbf{S}_M = \mathbf{D}_1\mathbf{S}_s\mathbf{D}_2^T\}$.

A general viewpoint to improve the optimization problem is that new information is added to that problem. The natural images are usually smooth. In this chapter, to improve image reconstruction performance, this information is used to limit the search space of the sparse recovery. Here, Wiener filter $WF(\mathbf{I}_\sigma, v_\sigma^2)$ is used to smooth the obtained image $\mathbf{I}_\sigma$ from the minimization of $J_\sigma(\mathbf{S}_s)$ with considering the noise power v_σ^2. In the proposed approach based on SL0 and Wiener filter, the following optimization steps for a fixed σ_k are proposed:

1.   Smoothing:

   a.   Obtaining the recovered image for a fixed σ_{k-1}:

$$\mathbf{I}_{\sigma_{k-1}} = \mathbf{\Psi} \times SCs^{-1}\{\mathbf{S}_s^{\sigma_{k-1}}\} \times \mathbf{\Psi^T}$$

   b.   Wiener filtering: $\hat{\mathbf{I}}_{\sigma_{k-1}} = WF(\mathbf{I}_{\sigma_{k-1}}, v_{\sigma_{k-1}}^2)$
   c.   Obtaining the scrambled sparse matrix as a starting point to minimize $J_{\sigma_k}(\mathbf{S}_s)$:

$$\mathbf{S}_{s0}^{\sigma_k} = SCc\{\mathbf{\Psi^T}\hat{\mathbf{I}}_{\sigma_{k-1}}\mathbf{\Psi}\}$$

2.   Optimizing the sparse measure $J_{\sigma_k}(\mathbf{S}_s)$ to obtain the scrambled sparse matrix $\mathbf{S}_s^{\sigma_k}$.

In this approach, the starting point to minimize $J_{\sigma_k}(\mathbf{S}_s)$ is smoothed via the Wiener filter. Before minimizing $J_{\sigma_k}(\mathbf{S}_s)$, the initial condition is projected on the subspace of the smoothed images by using the Wiener filter. In this chapter, the pixel-wise adaptive Wiener method based on statistics estimated from a local neighborhood of each pixel is used to smooth the estimated image $\mathbf{I}_{\sigma_{k-1}}$ as follows:

$$\hat{\mathbf{I}}_{\sigma_{k-1}}(x, y) = \mu_L + \frac{\sigma_L^2 - v_{\sigma_{k-1}}^2}{\sigma_L^2}\left(\mathbf{I}_{\sigma_{k-1}}(x, y) - \mu_L\right)$$

where μ_L and σ_L^2 are the image mean and standard deviation in the local neighborhood of the pixel (x, y). Here, the smoothness is also controlled by the parameter of noise power v_σ^2. In the proposed approach, a decreasing sequence of v_σ^2 is considered to control the smoothness by progressing the optimization of sparse recovery. When $\sigma \to 0$, the smoothing parameter v_σ^2 is approaching zero. In the proposed approach, the search space is limited by smoothing the initial solution of each step in the SL0 algorithm to escape from trapping into local minima. Simulation results show that this idea improves image reconstruction performance compared to the basic SL0 in this application.

Algorithm 1 illustrates the image decryption approach based on SL0 and Wiener filter. In this approach, inspired by [89], the noise effect is considered when the constraint $\|\mathbf{S}_M = \mathbf{D}_1\mathbf{S}_s\mathbf{D}_2^T\|_F \le \epsilon$ is not satisfied. In fact, by this condition, the constraint $\mathbf{S}_M = \mathbf{D}_1\mathbf{S}_s\mathbf{D}_2^T$ is relaxed to $\|\mathbf{S}_M - \mathbf{D}_1\mathbf{S}_s\mathbf{D}_2^T\|_F \le \epsilon$. The parameter ϵ depends on the noise level. The parameters are empirically used as follows: $L = 5$, $\mu = 0.5, \epsilon = 0.001$. The parameter $\sigma = [\sigma_1, \ldots, \sigma_J]$ is selected as $\sigma_1 = \max(|\hat{\mathbf{S}}_{s0}|)$, and $\sigma_j = \sigma_1(1 - 0.005j)$ which is a geometric decreasing approach. The decreasing sequence $v = [v_0, \ldots, v_{J-1}]$ is selected as 0.92^j.

Algorithm 1 The proposed decryption process based on SL0 and Wiener filter;

Inputs: $S_M, D_1, D_2, CSc\{.\}, CSc^{-1}\{.\}$
Outputs: Reconstructed image $\hat{I}$
Initialization:
(1) Choose two suitable decreasing sequences for σ, $[\sigma_1, \dots, \sigma_J]$ and $v = [v_0, \dots, v_{J-1}]$
(2) Initialization: $\hat{S}_{s0} = D_1^T(D_1 D_1^T)^{-1} S_M \left(D_2^T(D_2 D_2^T)^{-1}\right)^T$

For $j = 1, \dots, J$
  (1) Smoothing:
    • Obtaining the recovered image: $I_{\sigma_{j-1}} = \Psi \times SCs^{-1}\{\hat{S}_{s(j-1)}\} \times \Psi^T$
    • Wiener Filter: $\hat{I}_{\sigma_{j-1}} = WF\left(I_{\sigma_{j-1}}, v_{j-1}^2\right)$
    • Obtaining the scrambled sparse matrix: $\hat{S}_{s0}^{\sigma_{j-1}} = SCc\{\Psi^T \hat{I}_{\sigma_{j-1}} \Psi\}$
  (2) $\sigma = \sigma_j$
  (3) Initialization: $S = \hat{S}_{s0}^{\sigma_{j-1}}$
  (4) For $l = 1 \dots L$ (loop L times)
    (a) Let $\Delta = [\delta_{ij}]$, where $\delta_{ij} = s_{ij} \, exp\left(-\frac{s_{ij}^2}{2\sigma^2}\right)$.
    (b) Let $S \leftarrow S - \mu\Delta$ (μ is a small positive value).
    (c) Projection S onto the feasible set $\{S | S_M = D_1 S D_2^T\}$
      • If $\left\| S_M - D_1 S D_2^T \right\|_F > \epsilon$
$$S \leftarrow S - D_1^T(D_1 D_1^T)^{-1}(D_1 S D_2^T - S_M)\left(D_2^T(D_2 D_2^T)^{-1}\right)^T$$
      End if
    End for
  (5) $\hat{S}_{s(j)} = S$.

End For
$\hat{S}_s = \hat{S}_{s(j)}$
$\hat{I} = \Psi \times SCs^{-1}\{\hat{S}_s\} \times \Psi^T$

---

5 Experimental Results

In this section, different viewpoints such as security and statistical analysis, robustness in the presence of noise and cropping attack, and compression performance are used to evaluate the proposed method. The results are obtained on a personal computer with Intel Core i7-10,700 CPU @ 2.9 GHz and 16 GB RAM. Six images shown in Fig. 5, including Lena, Cameraman, Pepper, Barbara, Baboon, and MRI with size 256×256, are chosen to evaluate the proposed approach in different experiments. The wavelet transform is used and the initial values of the proposed

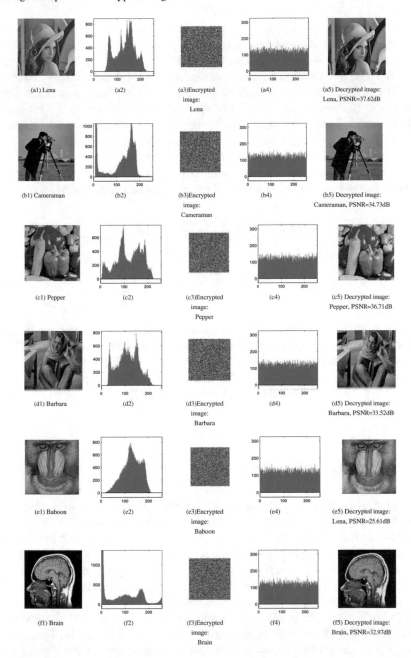

Fig. 5 Experimental results of compression-encryption based on 2D compressed sensing (CE-2DCS): (a1), (b1), (c1), (d1), (e1) and (f1) original images; (a2), (b2), (c2), (d2), (e2) and (f2) the histograms of the original images; (a3), (b3), (c3), (d3), (e3) and (f3) the encrypted images of the original images; (a4), (b4), (c4), (d4), (e4) and (f4) the histograms of the encrypted images; (a5), (b5), (c5), (d5), (e5) and (f5) the corresponding decrypted images

chaotic oscillator are $[x_{01}, y_{01}, z_{01}] = [-40, -2.85, -3.22]$, $[x_{02}, y_{02}, z_{02}] = [-39.9, -2.35, -3.52]$, and $[x_{03}, y_{03}, z_{03}] = [-39.77, -2.47, 2.76]$ respectively. The chaotic system is solved using the ODE45 MATLAB function and constant time step 0.01. It is appropriate to have different keystreams for various images to increase security and robustness against various attacks, such as the chosen-plaintext attack. It results in different measurement matrices and chaotic scrambling operators for various images. The dependency between the original image and keystream can be obtained from different approaches, such as generating the chaotic oscillator's initial value via the SHA-256 hash value of the input image [90].

Here, the performance of image reconstruction is assessed by two measures of the peak signal-to-noise ratio PSNR and the mean square deviation MSE as:

$$MSE = \frac{1}{n_1 n_2} \sum_{i,j} \left(I(i, j) - \hat{I}(i, j) \right)^2$$

$$PSNR = 10\log_{10}\left(\frac{255^2}{MSE} \right) \tag{6}$$

where $I \in R^{n_1 \times n_2}$ and $\hat{I} \in R^{n_1 \times n_2}$ are the original image and the decrypted image, respectively. Figures 5 and 6 show the performance of two schemes of CE-1DCS and CE-2DCS, respectively. These results show that all the encrypted images are meaningless with a uniform distribution. In this experiment, the compression ratio is set to 0.5. The proposed decryption process obtains an acceptable image reconstruction performance.

5.1 Compression Performance

An important measure to validate the performance of the encryption-compression process is the quality of the decrypted image for the different CRs. The Wiener filter is used in sparse recovery based on the SL0 algorithm in the sparse recovery and decryption process. Figure 7 presents the effect of the Wiener filter on the decrypted image in two schemes of CE-1DCS and CE-2DCS. The presented results show that the Wiener filter improves the PSNR values, approximately about 2 dB. The combination of the Wiener filter and SL0 algorithms increases the performance of the CE-1DCS approach significantly for small compression ratios. These results also show that CE-2DCS has better performance than CE-1DCS. Here, Fig. 8 presents the average computation time of the proposed reconstruction approach compared to the SL0 approach for 1D and 2D compression. In the rest of the chapter, the CE-2DCS scheme is used to evaluate and analyze the proposed approach. This comparison shows that the Wiener filter increases the computation time by approximately 3 s in this application with the mentioned hardware. Note that time is not appropriate to evaluate the computational complexity.

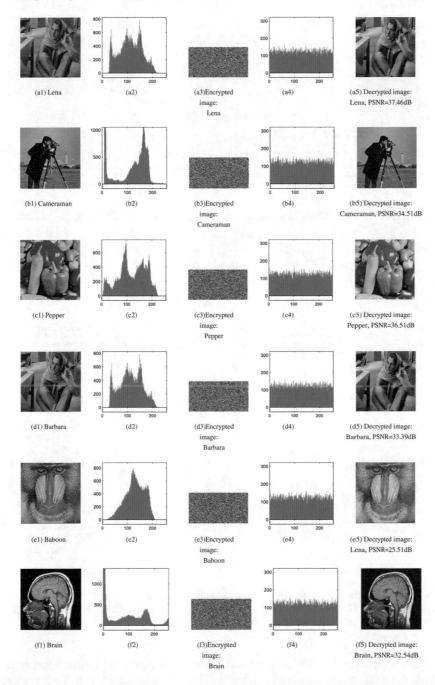

Fig. 6 Experimental results of compression-encryption based on 1D compressed sensing (CE-1DCS): (a1), (b1), (c1), (d1), (e1) and (f1) original images; (a2), (b2), (c2), (d2), (e2) and (f2) the histograms of the original images; (a3), (b3), (c3), (d3), (e3) and (f3) the encrypted images of the original images; (a4), (b4), (c4), (d4), (e4) and (f4) the histograms of the encrypted images; (a5), (b5), (c5), (d5), (e5) and (f5) the decrypted images

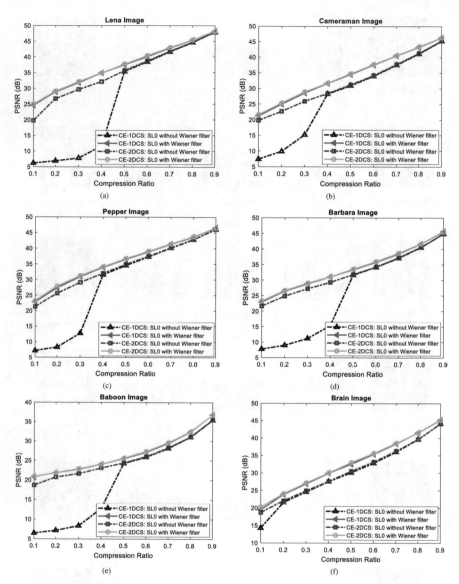

Fig. 7 Performance of the image reconstruction for two different compression-encryption approaches of CE-1DCS and CE-2DCS with considering the effects of Wiener filter in the sparse recovery: The PSNRs of the decrypted images for various CRs

Here, the performance of the proposed method is compared with other approaches based on compressed sensing in Table 1. These results depict that the proposed approach has better performance than compression-encryption methods [53, 57, 61, 63]. Table 2 shows the decrypted image quality visually. It can be seen that limiting the

Fig. 8 The calculation times in the various compression ratios

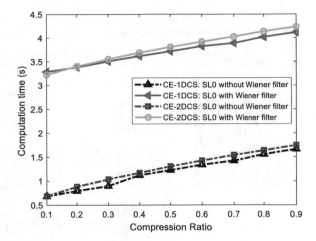

Table 1 PSNRs (dB) of the encrypted images

Image	Compression ratio	Proposed method	Hu et al. [61]	Xu et al. [53]	Chai et al. [63]	Hu et al. [57]
Lena	0.25	30.57	30.77	26.78	26.06	<20
	0.50	37.46	33.38	32.10	29.82	≈25.70
	0.75	44.23	32.28	32.35	29.56	≈33
Cameraman	0.25	27.14	–	24.24	25.23	<18
	0.50	34.73	–	30.66	29.43	≈23.50
	0.75	42.04	–	31.08	28.93	≈31
Pepper	0.25	29.55	30.56	–	–	<19
	0.50	36.71	33.73	32.09	–	≈26
	0.75	42.52	32.41	–	–	≈32.5

search space via the Wiener filter in the proposed sparse recovery obtains acceptable performance. It outperforms other encryption schemes.

5.2 Security Performance

Two key sensitivity and keyspace measures are used to analyze the proposed method's security performance based on compressed sensing and the proposed chaotic oscillator.

(a) **Key sensitivity**

An important measure showing the security level of a cryptosystem is key sensitivity. On the other hand, the decryption method cannot obtain the main image when there is

Table 2 The results of different compression ratios for the CE-2DCS approach; The first row and the second row related to each original image presents the encrypted and decrypted images, respectively

Original Image	CR = 0.1	CR = 0.25	CR = 0.75
	PSNR = 24.95dB	PSNR = 30.59dB	PSNR = 44.16dB
	PSNR = 21.61dB	PSNR = 27.20dB	PSNR = 41.94dB
	PSNR = 20.24dB	PSNR = 30.54dB	PSNR = 39.93dB

a tiny variation in the secret keys. Since the chaotic oscillator is sensitive to the initial values, our method is sensitive to a tiny secret key change. Here, the decrypted images of Lena are shown in the presence of a slight change of keystream in Fig. 9. In this experiment, the slight error of 10^{-14} is added to one key, and the others are correct. The obtained images are meaningless and entirely different from the original image. Figure 10 illustrates the results of the decrypted image based on MSE for a slight variation of each secret key. The proposed approach is sensitive to minor variations in the keystream.

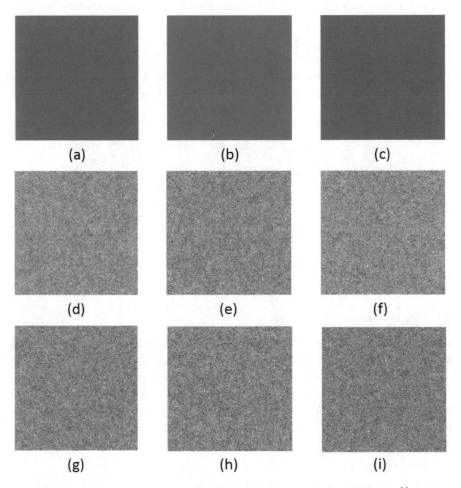

Fig. 9 Decrypted Lena image with incorrect keys; decrypted image with, **a** $x_{01} + 10^{-14}$; **b** $y_{01} + 10^{-14}$; **c** $z_{01} + 10^{-14}$; **d** $x_{02} + 10^{-14}$; **e** $y_{02} + 10^{-14}$; **f** $z_{02} + 10^{-14}$; **g** $x_{03} + 10^{-14}$; **h** $y_{03} + 10^{-14}$; **i** $z_{03} + 10^{-14}$

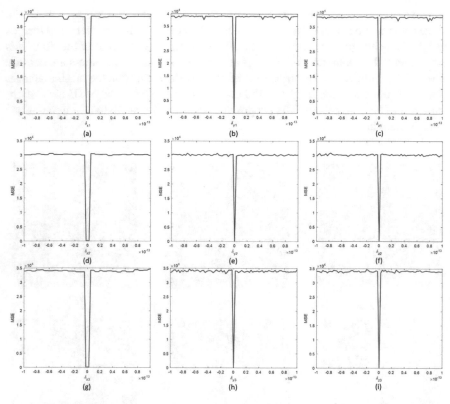

Fig. 10 MSE curves by changing each key: **a** $x_{01} + \delta_{x1}$; **b** $y_{01} + \delta_{y1}$; **c** $z_{01} + \delta_{z1}$; **d** $x_{02} + \delta_{x2}$; **e** $y_{02} + \delta_{y2}$; **f** $z_{02} + \delta_{z2}$; **g** $x_{03} + \delta_{x3}$; **h** $y_{03} + \delta_{y3}$; **i** $z_{03} + \delta_{z3}$

(b)   Keyspace

An attacker can use an approach to recover the original image. It is the brute force attack, an exhaustive search in the feasible set of keys. If the keyspace is larger than 2^{100}, this attack is infeasible [91]. Considering the data precision of 10^{-15} and 9 initial conditions of three used chaotic oscillators, the proposed scheme's keyspace is equal to $\left(10^{15}\right)^9 = 10^{135} \approx 2^{449}$. Hence, the proposed approach is sufficiently robust against the brute force attack. Here, the keyspace is also compared with other approaches in Table 3.

Table 3 Comparison of keyspace in different approaches

Algorithm	Proposed method	Hu et al. [61]	Xu et al. [53]	Chai et al. [63]	Hu et al. [57]
Keyspace	2^{449}	2^{449}	2^{398}	2^{232}	2^{200}

5.3 Statistical Study

Here, the security of the approach is analyzed via statistical measures.

(a) **Histogram**

Encryption approaches usually convert the original image to a meaningless image with a uniform distribution. The encrypted images are similar visually and statistically. So, the encryption scheme is robust to the statistical attack. The presented histograms in Figs. 5 and 6 show that the encryption process obtains approximately uniform distribution for all test images. Hence, the proposed cryptosystem has an acceptable performance using the Histogram analysis.

(b) **Entropy**

Another measure for the randomness of the meaningless image is information entropy. The security level is increased by maximizing the randomness and entropy. The maximum entropy of $\log_2 M$ where M is the number of discrete variable states, is obtained for the uniform distribution (the histogram flatness). In the proposed cryptosystem, 8 bit is considered for each pixel; hence the maximum entropy will be 8. The entropy of the original and encrypted images for $CR = 0.5$ are presented in Table 4. The proposed method approximately obtains the maximum randomness via the entropy viewpoint.

(c) **Correlation**

Two measures of histogram and entropy only consider the randomness via the random variable. The encrypted image is a random process, and the correlation among pixels can provide some information about the original image and encryption process. Hence, one measure for the irregularity in the encrypted image is the correlation among pixels. The security level of a cryptosystem is increased by minimizing the correlation. 2D histogram of adjacent pixels is an approach to represent the correlation between pixels visually. Figure 11 shows 2D histograms in the horizontal direction for the original and encrypted images. The proposed cryptosystem reduces the correlation between adjacent pixels. Here, the correlation coefficient (CC) is also used to quantify this correlation in three horizontal, vertical, and diagonal directions. Table 5 compares the CC values of the proposed scheme with other cryptosys-

Table 4 The entropy of the original and encrypted images

	Original image	Encrypted image
Lena	7.636492	7.994243
Cameraman	7.009716	7.994214
Pepper	7.772050	7.994234
Barbara	7.670796	7.994958
Baboon	7.570316	7.995191
Brain	6.986820	7.994585

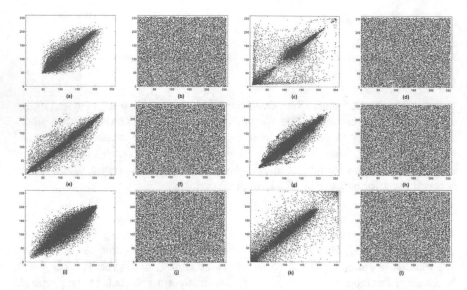

Fig. 11 The correlation in the horizontal direction for the main and encrypted pictures. **a** Main "Lena"; **b** encrypted "Lena"; **c** main "Cameraman"; **d** encrypted "Cameraman"; **e** main "Pepper"; **f** encrypted "Pepper"; **g** main "Barbara"; **h** encrypted "Barbara"; **i** main "Baboon"; **j** encrypted "Baboon"; **k** main "Brain"; **l** encrypted "Brain"

Table 5 Correlation coefficients of adjacent pixels

Image	Algorithms	Horizontal	Vertical	Diagonal
Lena	Original image	0.9721	0.9459	0.9212
	Proposed method	0.0029	−0.0051	0.0006
	Hu et al. [61]	−0.0046	−0.0002	0.0005
	Xu et al. [53]	0.0064	0.0003	0.0026
	Hu et al. [57]	0.0036	0.0012	0.0005
Cameraman	Original image	0.9592	0.9337	0.9079
	Proposed method	0.0013	−0.0059	−0.0030
	Hu et al. [61]	–	–	–
	Xu et al. [53]	0.0040	−0.0027	−0.0084
	Hu et al. [57]	–	–	–
Pepper	Original image	0.9714	0.9644	0.9388
	Proposed method	0.0012	0.0006	0.0041
	Hu et al. [61]	0.0003	0.0011	−0.0030
	Xu et al. [53]	−0.0117	0.0039	−0.0012
	Hu et al. [57]	0.0024	0.0057	0.0192

tems. The presented results show the acceptable performance of our approach in comparison with other approaches.

5.4 Robustness Analysis

(a) Noise attack

The transmission process usually disturbs the data by noise. In this subsection, the results of our method are evaluated in the presence of three noise models of Gaussian noise (GN), speckle noise (SN), and salt and pepper noise (SPN). The Lena picture with the compression ratio of 0.5 is used. Figure 12 presents the performance of the proposed cryptosystem for three noise models and different intensities. Figure 13 shows the effect of noise on the quality of the decrypted image. GN and SPN have the largest and the lowest effect. The proposed approach has certain robustness in the presence of the SN attack. These results show an acceptable performance of the cryptosystem in noise attack.

(b) Cropping attack

In this subsection, the performance of the approach is investigated with a cropping attack. In this attack, some pixels are lost. The proposed method is based on compressed sensing that measures the data structure via the measurement matrices. In this approach, each pixel of the encrypted image has some information about all original image pixels. Hence, our approach is robust against the cropping attack, and

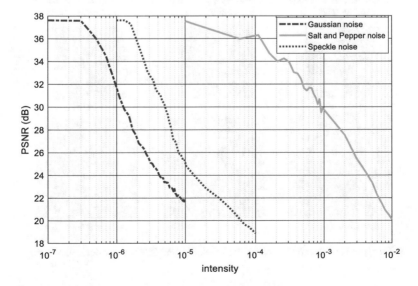

Fig. 12 PSNR of the decrypted and main images for various noises

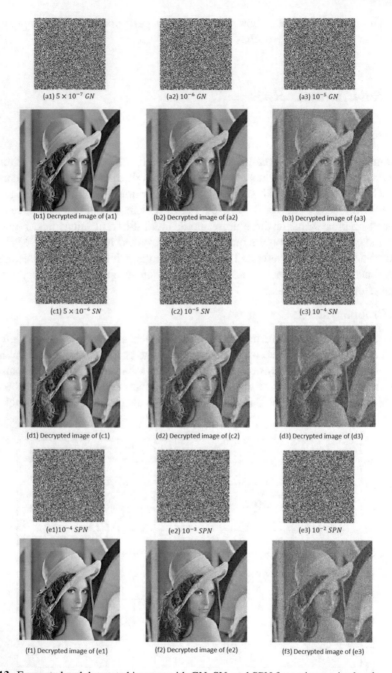

Fig. 13 Encrypted and decrypted images with GN, SN, and SPN for various noise levels

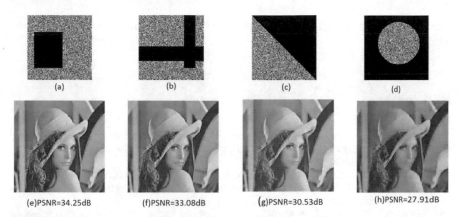

Fig. 14 Effect of cropping attacks: **a–d** encrypted images for various cropping attacks; **e–h** the decrypted images of (**a**)–(**d**)

the original image can be recovered via the sparse recovery. Here, different cropping masks shown in Fig. 14 are used to evaluate the encryption approach. The presented results in Fig. 14 demonstrate the effectiveness of the proposed approach with a severe cropping attack.

6 Conclusion

This chapter has presented an image compression-encryption approach using a novel chaotic oscillator and compressed sensing. First, the proposed chaotic oscillator was investigated and analyzed. This chaotic oscillator was used to construct two measurement matrices and chaotic scrambling operators. Two measurement matrices were applied to compress the original image based on the compressed sensing theory. In the proposed cryptosystem, the security level was increased by applying the chaotic scrambling and XOR operation with a random sequence generated by the chaotic sequence. Security analysis has shown that the proposed approach has high-level security concerning different viewpoints such as large keyspace, high sensitivity to the tiny variations of keys, uniform distribution of the encrypted image with maximum entropy, and weak correlation of adjacent pixels. These properties have described the robustness of the proposed cryptosystem against different attacks, such as brute-force attacks and statistical analysis. In the decryption process, the sparse recovery problem was solved via the smoothed ℓ_0 norm, a fast and accurate algorithm. In the proposed decryption, the Wiener filter was combined with the SL0 algorithm to reduce the search space and increase the compression performance. The simulation results showed the robustness of the proposed approach in the presence of noise and cropping attacks. The proposed algorithm has some advantages, such as a high security level and robustness against various attacks like cropping attacks. However, the computational complexity of the proposed method should be reduced in future works.

References

1. Abd El-Latif, A.A., Abd-El-Atty, B., Amin, M., Iliyasu, A.M.:Quantum-inspired cascaded discrete-time quantum walks with induced chaotic dynamics and cryptographic applications. Sci. Rep. **10**, 1–16 (2020)
2. Belazi, A., Abd El-Latif, A.A., Diaconu, A.-V., Rhouma, R., Belghith, S.:Chaos-based partial image encryption scheme based on linear fractional and lifting wavelet transforms. Opt. Lasers Eng. **88**, 37–50 (2017)
3. Belazi, A., Abd El-Latif, A.A., Belghith, S.:A novel image encryption scheme based on substitution-permutation network and chaos, Signal Process. **128**:155–170 (2016)
4. Pham, V.-T., Vaidyanathan, S., Volos, C., Jafari, S., Alsaadi, F.E.: Chaos in a simple snap system with only one nonlinearity, its adaptive control and real circuit design. Arch. Control Sci. **29** (2019)
5. Volos, C.K., Pham, V.-T., Nistazakis, H.E., Stouboulos, I.N.: A dream that has come true: chaos from a nonlinear circuit with a real memristor. Int. J. Bifurc. Chaos **30**, 2030036 (2020)
6. Nazarimehr, F., Pham, V.-T., Rajagopal, K., Alsaadi, F.E., Hayat, T., Jafari, S.: A new imprisoned strange attractor. Int. J. Bifurc. Chaos **29**, 1950181 (2019)
7. Pham, V.-T., Akgul, A., Volos, C., Jafari, S., Kapitaniak, T.: Dynamics and circuit realization of a no-equilibrium chaotic system with a boostable variable. AEU-Int. J. Electron. Commun. **78**, 134–140 (2017)
8. Pham, V.-T., Ouannas, A., Volos, C., Kapitaniak, T.: A simple fractional-order chaotic system without equilibrium and its synchronization. AEU-Int J Electron Commun **86**, 69–76 (2018)
9. Munoz-Pacheco, J.M., Zambrano-Serrano, E., Volos, C., Jafari, S., Kengne, J., Rajagopal, K.: A new fractional-order chaotic system with different families of hidden and self-excited attractors. Entropy **20**, 564 (2018)
10. Karthikeyan, A., Rajagopal, K., Kamdoum Tamba, V., Adam, G., Srinivasan, A.:A simple chaotic wien bridge oscillator with a fractional-order memristor and its combination synchronization for efficient antiattack capability. Complexity **2021** (2021)
11. Rajagopal, K., Singh, J.P., Karthikeyan, A., Roy, B.K.: Existence of metastable, hyperchaos, line of equilibria and self-excited attractors in a new hyperjerk oscillator. Int. J. Bifurc. Chaos **30**, 2030037 (2020)
12. Sambas, A., Vaidyanathan, S., Tlelo-Cuautle, E., Abd-El-Atty, B., Abd El-Latif, A.A., Guillén-Fernández, O., et al.: A 3-D multi-stable system with a peanut-shaped equilibrium curve: circuit design, FPGA realization, and an application to image encryption. IEEE Access **8**, 137116–137132 (2020)
13. Jafari, S., Sprott, J., Nazarimehr, F.: Recent new examples of hidden attractors. Eur. Phys. J. Spec. Top. **224**, 1469–1476 (2015)
14. Wang, R., Li, C., Çiçek, S., Rajagopal, K., Zhang, X.:A Memristive hyperjerk chaotic system: amplitude control, FPGA design, and prediction with artificial neural network. Complexity **2021** (2021)
15. Zhang, X., Li, C., Chen, Y., Herbert, H., Lei, T.: A memristive chaotic oscillator with controllable amplitude and frequency. Chaos Solitons Fractals **139**, 110000 (2020)
16. Sun, J., Li, C., Lu, T., Akgul, A., Min, F.: A memristive chaotic system with hypermultistability and its application in image encryption. IEEE Access **8**, 139289–139298 (2020)
17. Yu, Y., Shi, M., Kang, H., Chen, M., Bao, B.;Hidden dynamics in a fractional-order memristive Hindmarsh—Rose model. Nonlinear Dyn. 1–16 (2020)
18. Bao, H., Zhu, D., Liu, W., Xu, Q., Chen, M., Bao, B.: Memristor synapse-based morris-lecar model: bifurcation analyses and FPGA-based validations for periodic and chaotic bursting/spiking firings. Int. J. Bifurc. Chaos **30**, 2050045 (2020)

19. Bao, B., Peol, M., Bao, H., Chen, M., Li, H., Chen, B.: No-argument memristive hyper-jerk system and its coexisting chaotic bubbles boosted by initial conditions. Chaos Solitons Fractals **144**, 110744 (2021)
20. Akgul, A., Kengne, J., Rajagopal K., Pham, V.-T., Varan, M., Karthikeyan, A., et al.: Simulation and experimental implementations of memcapacitor based multi-stable chaotic oscillator and its dynamical analysis. Phys. Script. **96**, 015209 (2020)
21. Vaidyanathan, S., Tlelo-Cuautle, E., Anand, P.G., Sambas, A., Guillén-Fernández, O., Zhang, S.: A new conservative chaotic dynamical system with lemniscate equilibrium, its circuit model and FPGA implementation. Int. J. Autom. Control **15**, 128–148 (2021)
22. Vaidyanathan, S., Tlelo-Cuautle, E., Sambas, A., Dolvis, L.G., Guillén-Fernández, O.: FPGA design and circuit implementation of a new four-dimensional multistable hyperchaotic system with coexisting attractors. Int. J. Comput. Appl. Technol. **64**, 223–234 (2020)
23. Vaidyanathan, S., Tlelo-Cuautle, E., Sambas, A., Dolvis, L.G., Guillén-Fernández, O.: A new four-dimensional two-scroll hyperchaos dynamical system with no rest point, bifurcation analysis, multi-stability, circuit simulation and FPGA design. Int. J. Comput. Appl. Technol. **63**, 147–159 (2020)
24. Silva-Juárez, A., Tlelo-Cuautle, E., de la Fraga, L.G., Li, R.: Optimization of the Kaplan-Yorke dimension in fractional-order chaotic oscillators by metaheuristics. Appl. Math. Comput. **394**, 125831 (2021)
25. Rajagopal, K., Kingni, S.T., Khalaf, A.J.M., Shekofteh, Y., Nazarimehr, F.: Coexistence of attractors in a simple chaotic oscillator with fractional-order-memristor component: analysis, FPGA implementation, chaos control and synchronization. Eur. Phys. J. Spec. Top. **228**, 2035–2051 (2019)
26. Rajagopal, K., Nazarimehr, F., Guessas, L., Karthikeyan, A., Srinivasan, A., Jafari, S.: Analysis, control and FPGA implementation of a fractional-order modified Shinriki circuit. J. Circ. Syst. Comput. **28**, 1950232 (2019)
27. Cao, W., Zhou, Y., Chen, C.P., Xia, L.: Medical image encryption using edge maps. Signal Process. **132**, 96 109 (2017)
28. Belazi, A., Talha, M., Kharbech, S., Xiang, W.: Novel medical image encryption scheme based on chaos and DNA encoding. IEEE access **7**, 36667–36681 (2019)
29. Hua, Z., Yi, S., Zhou, Y.: Medical image encryption using high-speed scrambling and pixel adaptive diffusion. Signal Process. **144**, 134–144 (2018)
30. Patel, K.D., Belani, S.: Image encryption using different techniques: a review. Int. J. Emerg. Technol. Adv. Eng. **1**, 30–34 (2011)
31. Kaur, M., Kumar, V.: A comprehensive review on image encryption techniques. Arch. Comput. Methods Eng. **27**, 15–43 (2020)
32. Chen, G., Mao, Y., Chui, C.K.: A symmetric image encryption scheme based on 3D chaotic cat maps. Chaos Solitons Fractals **21**, 749–761 (2004)
33. Liu, H., Kadir, A.:Asymmetric color image encryption scheme using 2D discrete-time map. Signal Process. **113**, 104–112 (2015)
34. Han, J., Park, C.-S., Ryu, D.-H., Kim, E.-S.: Optical image encryption based on XOR operations. Opt. Eng. **38**, 47–54 (1999)
35. Ahmad, J., Khan, M.A., Ahmed, F., Khan, J.S.: A novel image encryption scheme based on orthogonal matrix, skew tent map, and XOR operation. Neural Comput. Appl. **30**, 3847–3857 (2018)
36. Zhou, R.-G., Wu, Q., Zhang, M.-Q., Shen, C.-Y.: Quantum image encryption and decryption algorithms based on quantum image geometric transformations. Int. J. Theor. Phys. **52**, 1802–1817 (2013)
37. Zhou, N.R., Hua, T.X., Gong, L.H., Pei, D.J., Liao, Q.H.: Quantum image encryption based on generalized Arnold transform and double random-phase encoding. Quantum Inf. Process. **14**, 1193–1213 (2015)
38. Zhang, Q., Guo, L., Wei, X.: Image encryption using DNA addition combining with chaotic maps. Math. Comput. Model. **52**, 2028–2035 (2010)

39. Wu, J., Liao, X., Yang, B.: Image encryption using 2D Hénon-Sine map and DNA approach. Signal Process. **153**, 11–23 (2018)
40. Telem, A.N., Fotsin, H.B., Kengne, J.: Image encryption algorithm based on dynamic DNA coding operations and 3D chaotic systems. Multimed. Tools Appl. 1–31 (2021)
41. Farah, M.B., Guesmi, A. Kachouri, R., Samet, M.: A novel chaos based optical image encryption using fractional Fourier transform and DNA sequence operation. Opt. Laser Technol. **121**, 105777 (2020)
42. Refregier, P., Javidi, B.: Optical image encryption based on input plane and fourier plane random encoding. Opt. Lett. **20**, 767–769 (1995)
43. Liu, S., Guo, C., Sheridan, J.T.: A review of optical image encryption techniques. Opt. Laser Technol. **57**, 327–342 (2014)
44. Moysis, L., Volos, C., Jafari, S., Munoz-Pacheco, J.M., Kengne, J., Rajagopal, K., et al.: Modification of the logistic map using fuzzy numbers with application to pseudorandom number generation and image encryption. Entropy **22**, 474 (2020)
45. Tsafack, N., Sankar, S., Abd-El-Atty, B., Kengne, J., Jithin, K., Belazi, A., et al.: A new chaotic map with dynamic analysis and encryption application in internet of health things. IEEE Access **8**, 137731–137744 (2020)
46. Sambas, A., Vaidyanathan, S., Tlelo-Cuautle, E., Abd-El-Atty, B., El-Latif, A.A.A., Guillén-Fernández, O., et al.: A 3-D multi-stable system with a peanut-shaped equilibrium curve: circuit design, FPGA realization, and an application to image encryption. IEEE Access **8**, 137116–137132 (2020)
47. Tsafack, N., Kengne, J., Abd-El-Atty, B., Iliyasu, A.M., Hirota, K., Abd El-Latif, A.A.: Design and implementation of a simple dynamical 4-D chaotic circuit with applications in image encryption. Inf. Sci. **515**, 191–217 (2020)
48. Ghaffari, A.: Image compression-encryption method based on two-dimensional sparse recovery and chaotic system. Sci. Rep. **11**, 1–19 (2021)
49. Alanezi, A., Abd-El-Atty, B., Kolivand, H., El-Latif, A., Ahmed, A., El-Rahiem, A., et al.: Securing digital images through simple permutation-substitution mechanism in cloud-based smart city environment. Secur. Commun. Netw. **2021** (2021)
50. Wang, H., Xiao, D., Li, M., Xiang, Y., Li, X.: A visually secure image encryption scheme based on parallel compressive sensing. Signal Process. **155**, 218–232 (2019)
51. Wen, W., Hong, Y., Fang, Y., Li M., Li, M..: A visually secure image encryption scheme based on semi-tensor product compressed sensing. Signal Process. **173**, 107580 (2020)
52. Yang, Y.-G., Wang, B.-P., Yang, Y.-L., Zhou, Y.-H., Shi, W.-M., Liao, X.: Visually meaningful image encryption based on universal embedding model. Inf. Sci. **562**, 304–324 (2021)
53. Xu, Q., Sun, K., He, S., Zhu, C.: An effective image encryption algorithm based on compressive sensing and 2D-SLIM. Opt. Lasers Eng. **134**, 106178 (2020)
54. Hua, Z., Zhang, K. Li, Y. Zhou, Y.: Visually secure image encryption using adaptive-thresholding sparsification and parallel compressive sensing. Signal Process. **183**, 107998 (2021)
55. Liu, X., Cao, Y., Lu, P., Lu, X., Li, Y.: Optical image encryption technique based on compressed sensing and Arnold transformation. Optik **124**, 6590–6593 (2013)
56. Zhou, N., Jiang, H., Gong, L., Xie, X.: Double-image compression and encryption algorithm based on co-sparse representation and random pixel exchanging. Opt. Lasers Eng. **110**, 72–79 (2018)
57. Hu, G., Xiao, D., Wang, Y., Xiang, T.: An image coding scheme using parallel compressive sensing for simultaneous compression-encryption applications. J. Vis. Commun. Image Represent. **44**, 116–127 (2017)
58. Zhu, L., Song, H., Zhang, X., Yan, M., Zhang, L., Yan, T.: A novel image encryption scheme based on nonuniform sampling in block compressive sensing. IEEE Access **7**, 22161–22174 (2019)
59. Chen, T., Zhang, M., Wu, J., Yuen, C., Tong, Y.: Image encryption and compression based on kronecker compressed sensing and elementary cellular automata scrambling. Opt. Laser Technol. **84**, 118–133 (2016)

60. Zhang, Y., Zhou, J., Chen, F., Zhang, L.Y., Wong, K.-W., He, X., et al.: Embedding cryptographic features in compressive sensing. Neurocomputing **205**, 472–480 (2016)
61. Hu, H., Cao, Y., Xu, J., Ma, C., Yan, H.: An image compression and encryption algorithm based on the fractional-order simplest chaotic circuit. IEEE Access **9**, 22141–22155 (2021)
62. Chen, J., Zhang, Y., Qi, L., Fu, C., Xu, L.: Exploiting chaos-based compressed sensing and cryptographic algorithm for image encryption and compression. Opt. Laser Technol. **99**, 238–248 (2018)
63. Chai, X., Zheng, X., Gan, Z., Han, D., Chen, Y.: An image encryption algorithm based on chaotic system and compressive sensing. Signal Process. **148**, 124–144 (2018)
64. Sun, C., Wang, E., Zhao, B.: Image encryption scheme with compressed sensing based on a new six-dimensional non-degenerate discrete hyperchaotic system and plaintext-related scrambling. Entropy **23**, 291 (2021)
65. Fan, J.-H., Liu, X.-B., Chen, Y.-B.: Image compression and encryption algorithm with wavelet-transform-based 2D compressive sensing. Opt. Appl. **49** (2019)
66. Zhou, N., Pan, S., Cheng, S., Zhou, Z.: Image compression—Encryption scheme based on hyper-chaotic system and 2D compressive sensing. Opt. Laser Technol. **82**, 121–133 (2016)
67. Deng, J., Zhao, S., Wang, Y., Wang, L., Wang, H., Sha, H.: Image compression-encryption scheme combining 2D compressive sensing with discrete fractional random transform. Multimed. Tools Appl. **76**, 10097–10117 (2017)
68. Ye, G., Pan, C., Dong, Y., Shi, Y., Huang, X.: Image encryption and hiding algorithm based on compressive sensing and random numbers insertion. Signal Process. 107563 (2020)
69. Gong, L., Deng, C., Pan, S., Zhou, N.: Image compression-encryption algorithms by combining hyper-chaotic system with discrete fractional random transform. Opt. Laser Technol. **103**, 48–58 (2018)
70. Chai, X., Fu, X., Gan, Z., Zhang, Y., Lu, Y., Chen, Y.: An efficient chaos-based image compression and encryption scheme using block compressive sensing and elementary cellular automata. Neural Comput. Appl. **32**, 4961–4988 (2020)
71. Chai, X., Gan, Z., Chen, Y., Zhang, Y.: A visually secure image encryption scheme based on compressive sensing. Signal Process. **134**, 35–51 (2017)
72. Ghaffari, A., Babaie-Zadeh, M., Jutten, C.: Sparse decomposition of two dimensional signals. In: 2009 IEEE international conference on acoustics, speech and signal processing, pp. 3157–3160 (2009)
73. Duarte, M.F., Baraniuk, R.G.: Kronecker compressive sensing. IEEE Trans. Image Process. **21**, 494–504 (2011)
74. Jokar, S., Mehrmann, V.: Sparse solutions to underdetermined Kronecker product systems. Linear Algebra Appl. **431**, 2437–2447 (2009)
75. Mohimani, H., Babaie-Zadeh, M., Jutten, C.: A fast approach for overcomplete sparse decomposition based on smoothed ℓ^0 norm. IEEE Trans. Signal Process. **57**, 289–301 (2008)
76. Donoho, D.L.: Compressed sensing. IEEE Trans. Inf. Theory **52**, 1289–1306 (2006)
77. Eldar, Y.C., Kutyniok, G.: Compressed sensing: theory and applications: Cambridge university press (2012)
78. Wright, J., Yang, A.Y., Ganesh, A., Sastry, S.S., Ma, Y.: Robust face recognition via sparse representation. IEEE Trans. Pattern Anal. Mach. Intell. **31**, 210–227 (2008)
79. Zhang, L., Yang, M., Feng, X.: Sparse representation or collaborative representation: which helps face recognition?. In: 2011 international conference on computer vision, pp 471–478 (2011)
80. Ghaffari, A., Fatemizadeh, E.: Sparse-induced similarity measure: mono-modal image registration via sparse-induced similarity measure. IET Image Proc. **8**, 728–741 (2014)
81. Ghaffari, A., Fatemizadeh, E.: Robust Huber similarity measure for image registration in the presence of spatially-varying intensity distortion. Signal Process. **109**, 54–68 (2015)
82. Aharon, M., Elad, M., Bruckstein, A.: K-SVD: an algorithm for designing overcomplete dictionaries for sparse representation. IEEE Trans. Signal Process. **54**, 4311–4322 (2006)
83. Zhang, Z., Chen, X., Liu, L., Li, Y., Deng, Y.: A sparse representation denoising algorithm for visible and infrared image based on orthogonal matching pursuit. SIViP **14**, 737–745 (2020)

84. Gribonval, R., Nielsen, M.: Sparse decompositions in unions of bases. IEEE Trans. Inform. Theory **49**, 3320–3325 (2003)
85. Eftekhari, A., Babaie-Zadeh, M., Moghaddam, H.A.: Two-dimensional random projection. Signal Process. **91**, 1589–1603 (2011)
86. Fang, Y., Wu, J., Huang, B.: 2D sparse signal recovery via 2D orthogonal matching pursuit. Sci. China Inf. Sci. **55**, 889–897 (2012)
87. Chen, G., Li, D., Zhang, J.: Iterative gradient projection algorithm for two-dimensional compressive sensing sparse image reconstruction. Signal Process. **104**, 15–26 (2014)
88. Blake, A., Zisserman, A.: Visual reconstruction. MIT press (1987)
89. Eftekhari, A., Babaie-Zadeh, M., Jutten, C., Moghaddam, H.A.: Robust-SL0 for stable sparse representation in noisy settings. In: 2009 IEEE international conference on acoustics, speech and signal processing, pp. 3433–3436 (2009)
90. Zhu, S., Zhu, C., Wang, W.: A novel image compression-encryption scheme based on chaos and compression sensing. IEEE Access **6**, 67095–67107 (2018)
91. Alvarez, G., Li, S.: Some basic cryptographic requirements for chaos-based cryptosystems. Int. J. Bifurc. Chaos **16**, 2129–2151 (2006)

Backstepping and Sliding Mode Control of a Fractional-Order Chaotic System

Akif Akgul, Murat Erhan Cimen, Muhammed Ali Pala,
Omer Faruk Akmese, Hakan Kor, and Ali Fuat Boz

Abstract Fractional-order systems can come across in chemical processes, biological systems, viscoelastic systems, propagation of electromagnetic waves and electrochemical systems. In the literature, fractional systems are encountered in the modelling and controlling of such systems, synchronization applications, communication, modelling and controlling of power systems or chemical processes. Many methods such as PID, sliding mode control, backstepping control, fuzzy sliding mode control, model predictive control, reinforcement learning and adaptive sliding mode control, have been used in the control of such systems. In this study, a fractional-order chaotic system is proposed. For this fractional-order chaotic system, bifurcation, phase portraits and Lyapunov exponents have been calculated to investigate its chaotic status. Then, in order to control the chaotic system, mathematically backstepping and sliding mode method control laws are obtained and their applications are realised. Consequently, their system responses by means of backstepping and sliding mode control results are discussed and compared with each other. Nevertheless, both controllers successfully are be able to regulate the system by obtained control laws, even if the system is out of the chaotic situation for specified fractional-order.

1 Introduction

Lorenz introduced the phenomenon of chaos and the butterfly attractor, which is very sensitive to initial conditions, in 1963 [1]. In recent years, the control and synchronization of chaotic systems has attracted the attention of researchers [2, 3]. Chaotic systems with uncertainty in theory emerge as a paradigm for synchronization problems in many studies because they are both simple for mathematical operations and can be used in practice [4, 5].

A. Akgul (✉) · O. F. Akmese · H. Kor
Department of Computer Engineering, Faculty of Engineering, Hitit University, Corum, Turkey
e-mail: akifakgul@hitit.edu.tr

M. E. Cimen · M. A. Pala · A. F. Boz
Department of Electrical and Electronics Engineering, Faculty of Technology, Sakarya University of Applied Sciences, Serdivan, Sakarya, Turkey

© The Author(s), under exclusive license to Springer Nature Switzerland AG 2022
A. A. Abd El-Latif and C. Volos (eds.), *Cybersecurity*, Studies in Big Data 102,
https://doi.org/10.1007/978-3-030-92166-8_3

Non-integer calculus was introduced by Leibniz in 1695. For nearly three centuries, fractional calculus theory has developed as a purely theoretical field of mathematics. For a long time, fractional calculus was considered the only mathematical and theoretical science with almost no application [6]. The fractional calculus, which is only used in mathematics, has not been discovered until recently in other sciences [3]. However, in recent years, great attention has been paid to the applications of fractional calculus in engineering and physical systems. A Fractional-order chaotic system, which is a generalization of integer-ordered chaotic system, is accepted as a new alternative for the development of techniques for modeling, synchronization and control of generalized dynamical systems. It has applications in many fields such as cryptography, secure communication, synchronization, random number generators, robotics, wave propagation and genetic algorithms [7–12]. One of the most striking applications is the fractional-order controller [13]. Fractional chaotic systems have more complex characteristic structures compared to other chaotic systems. This has significantly increased good stability and tracking ability in controlling complex systems. For these reasons, a wide application area has emerged in the control theory of fractional chaotic systems, such as adaptive control, linear error feedback control, active pinning control, impulsive control, fuzzy logic controller [14–16]. Also many researchers have contributed to smart control approaches ranging from traditional to advanced technology [17, 18].

Fractional-order chaotic systems are considered as generalizations of integer-ordered chaotic systems, alternatives for which considerable attention has been motivated to develop modelling methods, controlling or synchronizing this class of generalized dynamical systems [18–20]. Fractional-order chaotic systems can exhibit much richer dynamic behaviour compared to integer-order chaotic systems and are therefore considered powerful tools in secure communication due to their ability to increase the security of chaotic communication systems [21]. A fractionally ordered chaotic system is asymmetrical, dissimilar, and not topologically equivalent to typical chaotic systems. It challenges the traditional view that the existence of unstable equilibria is necessary to ensure the existence of chaos [22].

As a novelty in this study, a fractional chaotic system is successfully controlled by making different fractional analyzes and comparing these parameters with two different methods for controlling the system. In this study, analyzes were carried out on the phase diagram of a chaotic fractional system, bifurcation diagram, and Lyapunov exponents in the literature. Then, for this fractional-order chaotic system, backstepping and sliding mode method control laws were obtained mathematically and their applications were performed. Fractional degree changing effect investigate on for both controllers. As a result, the obtained results and system responses are discussed and compared with each other. In addition, although the system comes out of the chaotic state, it can be successfully regulated by the control laws obtained in both controllers.

2   The Used Fractional-Order Chaotic System

A fractional-order system can be given as in Eq. 1. Here $\propto$ [0, 1] denotes fractional degree. C is continuous, t_0 is the start time, t is time, $x(t)$ is the value of an orbit concerning time. D stands for the derivative operator.

$$\, _{t_0}^C D_t^\alpha x(t) = f(x, \ t) \tag{1}$$

Equation 1 from the equation in lemma 2 given in [14].

$$\frac{1}{2}\, _{t_0}^C D_t^\alpha x^2(t) \le x(t)\, _{t_0}^C D_t^\alpha x(t) \tag{2}$$

The used chaotic system in the paper is given in Eqs. 3 and 4 as integer and fractional-order forms. Initial conditions of the chaotic system are [0.25 0.25 0.25] for x, y, z states.

$$\begin{aligned} \dot{x} & \quad -ax + by + yz \\ \dot{y} & = -ay - bx + xz \\ \dot{z} & \quad +z - xy + u \end{aligned} \tag{3}$$

$$\begin{aligned} _0^C D_t^\alpha x & \quad -ax + by + yz \\ _0^C D_t^\alpha y & = -ay - bx + xz \\ _0^C D_t^\alpha z & \quad +z - xy + u \end{aligned} \tag{4}$$

Analyzes such as Poincare map, bifurcation diagrams, time series, phase portrait and Lyapunov exponents are used to show that a system is chaotic. In this study, the bifurcation diagrams, phase portraits and Lyapunov exponents of the fractional-order system are examined and the situations in which chaos occurs are examined.

2.1   Bifurcation Diagrams

The variation of the parameter-dependent states of a system with respect to time can be shown with bifurcation diagrams. In the bifurcation diagram given in Fig. 1.a, while the fractional degree ($\propto$) 0.95 of the chaotic system in Eq. 4 for $u = 0$, bifurcation diagram showing only the values in which the x state takes the parameter in the range $0 \le a \le 4$ is drawn. In Fig. 1b for same system, $u = 0$, fractional-order, according to ($\propto$) 0.95, $a = 2\ 0 \le a \le 5$, the bifurcation diagram showing the values taken by the x state is given.

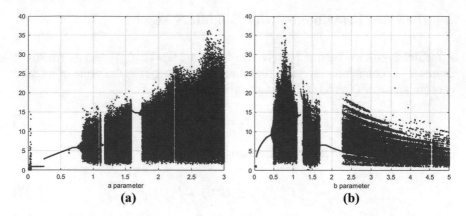

Fig. 1 Bifurcation diagrams for the used fractional-order chaotic system, **a** the bifurcation graph of state x for varying parameters a, $\propto = 0.95$, $b = 1$. **b** The bifurcation graph of state x for varying parameters b, $\propto = 0.95$, $a = 2$

2.2 Phase Portraits

It can be seen by the phase portraits that the trajectories of a system are displayed relative to each other. Phase portraits can also provide the display of the behaviour of the system around the equilibrium point of the system. In this study, while the fraction order $(\propto) = 1, 0.999, 0.98, 0.95, 0.9$ and 0.85 of the used system for $u = 1$, the phase diagrams of the $x - y$ state are given in Fig. 2a–f.

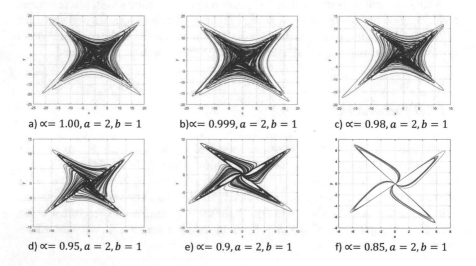

a) $\propto = 1.00, a = 2, b = 1$ b) $\propto = 0.999, a = 2, b = 1$ c) $\propto = 0.98, a = 2, b = 1$

d) $\propto = 0.95, a = 2, b = 1$ e) $\propto = 0.9, a = 2, b = 1$ f) $\propto = 0.85, a = 2, b = 1$

Fig. 2 Phase portraits of the used fractional-order chatic system

2.3 Lyapunov Exponents

Lyapunov exponentials are numerical quantities that show the sensitivity of the system relative to its initial state. In other words, Lyapunov exponentials measure the mean rate of convergence or divergence of trajectories or orbits from the initial state. Thus, they can be used to analyse the stability of the system, and furthermore, they can be used to analyse whether time series contains chaotic attractor or not. Calculation of Lyapunov exponents using Gram Schmidt orthogonalization method in integer-order continuous systems was firstly [23]. This method was improved and other alternative methods are developed that based on system derivative and scalar product of system's perturbation [24–26]. However, this method was not suitable for discontinuous systems. Studies have been carried out to calculate Lyapunov exponents in discontinuous and non-smooth integer-order systems [Determining Lyapunov exponents of non-smooth systems: Perturbation vectors approach, Calculation of Lyapunov exponents for dynamic systems with discontinuities, Using chaos synchronization to estimate the largest Lyapunov exponent of nonsmooth systems, Estimation of the largest Lyapunov exponent in systems with impacts Evaluation of the largest Lyapunov exponent in dynamical systems with time delay]. Li et al., on the other hand, carried out work on calculating Lyapunov exponents in fractional-order systems [On the bound of the Lyapunov exponents for the fractional differential systems]. Danca and Kuznetsov prepared a Matlab Program that employs Adams-Bashforth-Moulton method for calculating the Lyapunov exponents of the fractional-order system [27].

For any system to be chaotic, one of the Lyapunov exponents must be positive. Otherwise, the system is not chaotic anyway. Since the used system is 3D, three Lyapunov exponents are calculated. If three Lyapunov exponents are negative $(-, -, -)$ than system is stable. If two Lyapunov exponents are negative and one is zero $(-, -, 0)$ than system is stable limit. If one Lyapunov exponent is negative and two Lyapunov exponents are zero $(-, 0, 0)$ than system is torus. If one Lyapunov exponent is positive, one Lyapunov exponent is negative and one Lyapunov exponent is zero $(+, -, 0)$ than system is stranger attractor, that is the system has chaotic behaviour. For the fractional-order system, the calculated result given in Fig. 3a is calculated as the fractional-order of the system $(\alpha = 0.95)$ and the Lyapunov exponents are $\lambda_1 = 0.5931$, $\lambda_2 = 0.009$ and $\lambda_3 = -4.3093\ 0$, that is strange attractor $(+, 0, -)$, for $a = 2, b = 1$. The fractional-order system is chaotic because of positive exponent. When $\alpha = 0.95$ and a $= 2$, all Lyapunov exponents for $0 \leq b \leq 5$ are calculated and given graphically in Fig. 3b. In Fig. 3c, when the fractional-order of the system is $\alpha = 0.95$ and b $= 1$, all Lyapunov exponents for $0 \leq a \leq 3$ are calculated and given graphically. Finally, in Fig. 3d, the parameters of the system are given graphically when a $= 2$ and b $= 1$ and all Lyapunov exponents are calculated for fractional-order $0.5 \leq \alpha \leq 1$.

The Lyapunov exponents obtained according to the different fractional-order of the used chaotic system are given in Table 1. When this table is examined, the

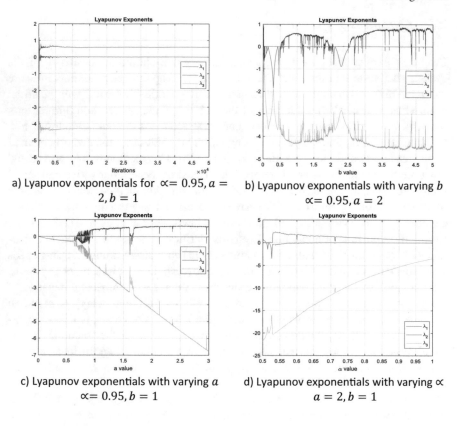

a) Lyapunov exponentials for $\propto= 0.95, a = 2, b = 1$

b) Lyapunov exponentials with varying b $\propto= 0.95, a = 2$

c) Lyapunov exponentials with varying a $\propto= 0.95, b = 1$

d) Lyapunov exponentials with varying $\propto$ $a = 2, b = 1$

Fig. 3 Lyapunov exponents for different parameter values

Table 1 Lyapunov exponents for different fractional-orders of the used chaotic system

Fractional-order	λ_1	λ_2	λ_3
0.999	0.4933	0.0030	−3.5085
0.995	0.4723	−0.0014	−3.5354
0.99	0.4812	−0.0001	−3.6121
0.98	0.5140	0.0006	−3.7830
0.95	0.5931	0.0009	−4.3093
0.90	0.6989	−0.0004	−5.2879
0.85	0.8540	0.0007	−6.5019
0.8	0.9983	−0.0001	−7.9230
0.75	1.1637	−0.0025	−9.6024
0.7	1.3955	−0.0085	−11.6137
0.6	0.6088	−0.0884	−15.4067
0.5	0.9829	−0.1112	−21.8866

system has chaotic behaviour in the given fractional-order values because of one of the Lyapunov exponents are positive.

3   The Controller Design for Fractional-Order Chaotic System

The general structure of fractional degree systems is given in Eq. 1. If the system of Eq. 1 is $x(t) \in \mathbb{R}$ for the equilibrium point $x = 0$, then the condition in Eq. 3 is satisfied [14]. If this system is at equilibrium and does not move, the system is stable. In addition, for the system to be asymptotically stable, it must satisfy Eq. 4 [28].

$$x(t)f(x(t), t) \leq 0 \; \forall x \tag{5}$$

$$x(t)f(x(t), t) < 0 \; \forall x \neq 0 \tag{6}$$

For an exemplary system, the Lyapunov energy function can be given as in Eq. 5. However, if the system fractional degree $\alpha = 1$, the particular case $\dot{V} = x\dot{x}$ can be obtained. This condition is not provided when the fractional degree is different. However, when Eqs. 2 and 5 are substituted, Eq. 6 can be obtained. This inequality can be used very easily for fractional-order systems. In this case, for the states of a sample system in Eq. 1 to be stable, the condition in Eq. 8 must be met. This expression, defined by the Lyapunov function, becomes a negative semidefinite because it is an inequality. In this case, according to the obtained expression in Eq. 7, the equilibrium point of the function becomes stable [28].

$$V(x(t), t) = \frac{1}{2}x^2(t) \tag{7}$$

$$_{0}^{C}D_{t}^{\alpha}V(x(t), t) = \frac{1}{2}{}_{0}^{C}D_{t}^{\alpha}x^2(t) \leq x(t){}_{0}^{C}D_{t}^{\alpha}x(t) \tag{8}$$

$$x_1{}_{0}^{C}D_{t}^{\alpha}x_1(t) \leq 0 \; \forall x_1 \tag{9}$$

In this context, using the expression obtained in Eq. 3, the sliding mode and backstepping methods of the control of the system given in Eq. 10 and controller designs will be realized.

3.1 Backstepping Control Design

Let the Lyapunov function for the system given in Eq. 9 be as in Eq. 10. Then, the approach given in Eq. 6 will be used to control it with the backstepping controller designed for the Fractional system. This approach is based on inequality as in [28]. Equation 11 is obtained by first taking the fractional derivatives. Then, the situations here are written in their places, and Eq. 12 and Eq. 13 are obtained. For the virtual controller design, the variable ζ is defined as in Eq. 14.

$$V = \frac{1}{2}x^2 + \frac{1}{2}y^2 \tag{10}$$

$$ {}_{0}^{C}D_{t}^{\alpha}V = \frac{1}{2}{}_{0}^{C}D_{t}^{\alpha}x^2 + \frac{1}{2}{}_{0}^{C}D_{t}^{\alpha}y^2 \leq x{}_{0}^{C}D_{t}^{\alpha}x + y{}_{0}^{C}D_{t}^{\alpha}y \tag{11}$$

$$ {}_{0}^{C}D_{t}^{\alpha}V = x(-ax + by + yz) + y(-ay - bx + xz) \tag{12}$$

$$ {}_{0}^{C}D_{t}^{\alpha}V = -ax^2 - ay^2 + 2xyz \tag{13}$$

$$\zeta = z_d = -yx \tag{14}$$

Equation 15 is obtained by substituting the virtual controller defined in Eq. 14 for Eq. 13. Here an e_1 error is determined as Eq. 16. The main goal is for it to converge to 0. In order to include the determined error in the Lyapunov equation, the z variable is left alone, as in Eq. 18. Equation 18 is written instead of z in Eq. 19 to get Eqs. 20 and Eq. 21 with its final form.

$$ {}_{0}^{C}D_{t}^{\alpha}V = -ax^2 - ay^2 + 2xyz \rightarrow {}_{0}^{C}D_{t}^{\alpha}V = -ax^2 - ay^2 - 2y^2x^2 \leq 0 \tag{15}$$

$$e_1 = z - \zeta = z - z_d = z + yx \tag{16}$$

$$e_1 = z + xy \tag{17}$$

$$z = e_1 - xy \tag{18}$$

$$ {}_{0}^{C}D_{t}^{\alpha}y = -ay - bx + xz \tag{19}$$

$$ {}_{0}^{C}D_{t}^{\alpha}y = -ay - bx + x(e_1 - xy) \tag{20}$$

$$ {}_{0}^{C}D_{t}^{\alpha}y = -ay - bx + xe_1 - x^2y \tag{21}$$

In order to follow the transactions, the e_1 error signal is given again in Eq. 22. From here on, the fractional derivative is retaken according to this e_1 variable. However, under normal circumstances, the lehospital rule does not apply to systems of fractional degrees [29]. For this, Eq. 22 given in [30, 31] must be provided.

$$\|D_t^\alpha (fg)\|_r \leq \|D_t^\alpha (f)\|_{p_1} \|g\|_{p_1} + \|D_t^\alpha (g)\|_{p_2} \|f\|_{p_2} \tag{22}$$

In this case, when e_1 is rewritten, Eq. 23 is obtained. Equation 24 is obtained when the fractional derivative of this equation is taken. Then a Lyapunov function is reconstructed in e_1. Equation 26 is obtained by taking the derivative of this Lyapunov function. Equation 27 is obtained when the system states $D_t^q x$ and $D_t^q y$ are written into the equation. Editing Eq. 27 yields Eqs. 28–30, and finally, Eq. 31. In order for Eq. 31 to be stable, it must be negative definite, that is, it must be equal to a value such as $2xy +_0^C D_t^\alpha e_1 = -ke_1$ to have $_0^C D_t^\alpha V \leq 0$.

$$e_1 = z + xy \tag{23}$$

$$
\begin{aligned}
_0^C D_t^\alpha e_1 &\neq D_t^q z + y D_t^q x + x_0^C D_t^\alpha y \;\rightarrow\; _0^C D_t^\alpha e_1 = D_t^q z \\
&+ \left\| D_t^q x \right\|_{p_1} \|y\|_{p_1} + \left\| D_t^q y \right\|_{p_2} \|x\|_{p_2}
\end{aligned}
\tag{24}
$$

$$V = x^2 + y^2 + e_1^2 \tag{25}$$

$$D_t^q V = x D_t^q x + y D_t^q y + e_1 {}_0^C D_t^\alpha e_1 \tag{26}$$

$$_0^C D_t^\alpha V = x(-ax + by + yz) + y(-ay - bx + xz) + e_1 {}_0^C D_t^\alpha e_1 \tag{27}$$

$$D_t^q V = -ax^2 - by^2 + 2xyz + e_{10}{}^C D_t^\alpha e_1 \tag{28}$$

$$_0^C D_t^\alpha V = -ax^2 - by^2 + 2xy(e_1 - xy) + e_1 {}_0^C D_t^\alpha e_1 \tag{29}$$

$$_0^C D_t^\alpha V = -ax^2 - by^2 - 2x^2 y^2 + 2xye_1 + e_1 {}_0^C D_t^\alpha e_1 \tag{30}$$

$$_0^C D_t^\alpha V = -ax^2 - by^2 - 2x^2 y^2 + e_1 \underbrace{\left(2xy +_0^C D_t^\alpha e_1\right)}_{-ke_1} \tag{31}$$

Equation 32 is obtained when Eq. 22 is applied to $_0^C D_t^\alpha e_1$ in Eq. 31. Equation 33 is obtained by substituting the system states in Eq. 32. When the control sign is taken from Eq. 33, the control sign is obtained in Eq. 34. In this controller, $k = 1$, $p_1 = p_2 = 2$ is set. The general view of this designed control structure is given in Fig. 4. This control signal is applied to the system after 100 s.

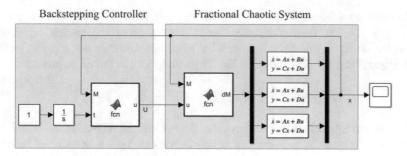

Fig. 4 Backstepping control structure of fractional-order system

$$
{}_0^C D_t^\alpha V = -ax^2 - by^2 - 2x^2y^2
$$
$$
+e_1\left(2xy + D_t^q z + \left\|D_t^q x\right\|_{p_1}\|y\|_{p_1} + \left\|D_t^q y\right\|_{p_2}\|x\|_{p_2}\right) \tag{32}
$$

$$
{}_0^C D_t^\alpha V = -ax^2 - by^2 - 2x^2y^2
$$
$$
+e_1\left(2xy + z - xy + u + \|-ax + by + yz\|_{p_1}\|y\|_{p_1} + \left\|-ay - bx + xe_1 - x^2y\right\|_{p_2}\|x\|_{p_2}\right) \tag{33}
$$

$$
u = -ke_1 - 2xy - z + xy + \|-ax + by + yz\|_{p_1}\|y\|_{p_1}
$$
$$
+\left\|-ay - bx + xe_1 - x^2y\right\|_{p_2}\|x\|_{p_2} \tag{34}
$$

The response of the system controlled by backstepping are given in Fig. 5. It can be seen in the system responses that the system can be controlled even when it is of different fractional-order.

3.2 Sliding Mode Control Design

There are two stages in the sliding mode control. The first is the determination of the surface on which the trajectories of the system will move, and the second is the determination of the control law that will make the switching. Here, the first aim is to carry the trajectories of the system at any operating point to this designed surface. Then, an infinite frequency of the system states carried to this surface is brought to equilibrium with a close switched control laws. The surface to be designed for a fractional-order system is given in Eq. 35 [32]. Again, the Lyapunov energy function is constructed according to the surface in Eq. 35 as in Eq. 36. For the Lyapunov function in Eq. 36 to be stable, its derivative must be negative. Equation 37 is obtained by taking the derivative of the Lyapunov function in Eq. 36. The derivative of the surface in Eq. 37 is given in Eq. 38.

In the logic here, the condition in Eq. 39 that will make a switching must be satisfied for $\dot{V} = s\dot{s} \le 0$ to be. k_3 is a positive constant greater than 0 [32]. When

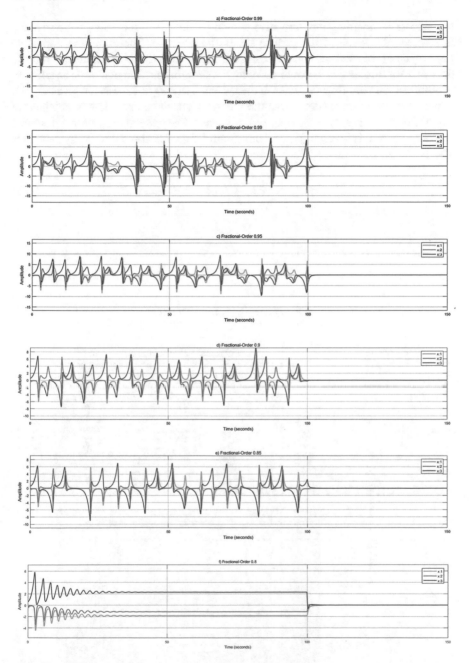

Fig. 5 Backstepping control after 100 s using different fractional values of the fractional-order system

the surface is positive, the derivative should be negative, and when the derivative is positive, the value of the surface should be negative. Therefore, the derivative expression of this surface can be expressed with a negative signum function. If the derivative of this surface is combined with the signum function of the surface, Eq. 40 is obtained. Substituting the states of the system in the $D_t^{\alpha} z$ expression gives Eq. 41. When u is left alone to determine the control sign, the control law is obtained in Eq. 42. In this study, $k_1 = k_2 = k_3 = 1$ are taken. The general view of this designed control structure is given in Fig. 6. This control signal is applied to the system after 100 s.

$$s = k_1 D_t^{\alpha-1} z + k_2 \int (c_1 x + c_2 y + c_3 z) dt \qquad (35)$$

$$V = s^2 \qquad (36)$$

$$\dot{V} = s\dot{s} \leq 0 \qquad (37)$$

$$\dot{s} = k_1 D_t^{\alpha} z + k_2 (c_1 x + c_2 y + c_3 z) \qquad (38)$$

Sliding Mode Controller   Fractional Chaotic System

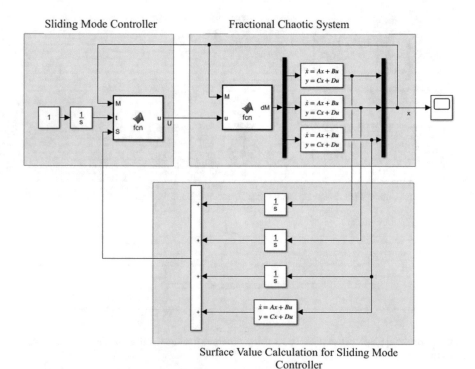

Surface Value Calculation for Sliding Mode
Controller

Fig. 6 Sliding mode control structure of a fractional-order system

$$\dot{s}(x) = \begin{cases} k_3, s < 0 \\ -k_3, s \geq 0 \end{cases} \tag{39}$$

$$\dot{s} = k_1 D_t^\alpha z + k_2(c_1 x + c_2 y + c_3 z) = -k_3 sgn(s) \tag{40}$$

$$k_1(z - xy + u) + k_2(c_1 x + c_2 y + c_3 z) = -k_3 sgn(s) \tag{41}$$

$$u = \frac{1}{k_1}(-k_3 sgn(s) - z + xy - k_2(c_1 x + c_2 y + c_3 z)) \tag{42}$$

When the obtained results are controlled, it is seen in Fig. 7 that the system can control the sliding mode even when the fractional degree changes. It can be seen that the system with a high fractional degree reaches the equilibrium point of 0. As can be seen in Fig. 7, when the fraction degree is 0.8, it is seen that the system response does not reach the equilibrium point. That is, it reaches a different equilibrium point. This is since the surface passes through this equilibrium point. Therefore, the surface design for the slide mode controller needs to be updated when it will be controlled with a low fractional sliding mode.

4   Discussion

The fractional-order chaotic system used in the study was analysed. The methods of bifurcation diagrams, phase portraits and later Lyapunov exponentials were used as the analysis of the Fractional Chaotic System. When the bifurcation diagrams are examined, it is seen that the chaotic state of the system changes according to the changes in the a, b or α parameters of the system for the fraction degree 0.95. When the phase portraits are examined, the phase portraits of the chaotic system are given for the different fraction degrees of the system against the constant a and b parameters. As can be seen, when the fraction degree decreases, it is seen that the behaviour of the system changes and the attractor shape of the system also changes. When the Lyapunov exponents are examined, the Lyapunov exponents of the fractional-order system take positive, zero or negative values according to the changing parameters. In this case, the attractor shape changes. Then, sliding mode and backstepping control methods were used to control the system. When the fraction degree of the system decreases due to the surface used in the sliding mode, the system is located at the stable point in the system. Therefore, for the system to reach the desired stable point, it is necessary to redesign a controller for the system when the fractional degree changes. However, there was no such problem in the backstepping control method and the system settled at the desired stable point. In addition, there is no chattering problem in the backstepping control method, since high-frequency

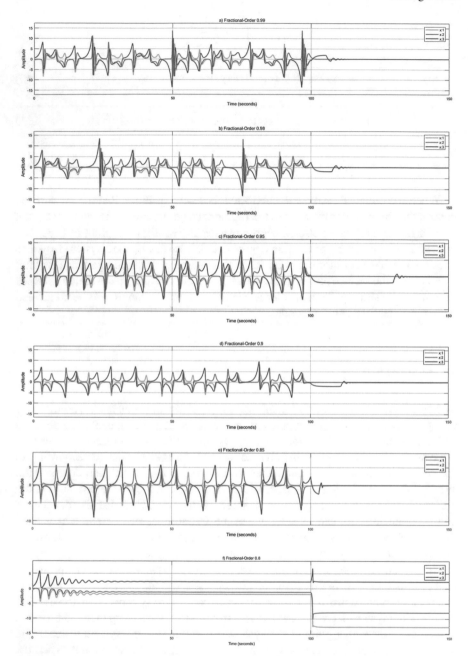

Fig. 7 Control of the fractional-order system with sliding mode after 100 s using different fraction values

switching is not performed. However, it is seen that the control method designed with the backstepping method reaches the stable point more quickly than the sliding mode method.

5 Conclusion

In this study, analyzes were carried out on the phase diagram of a fractional-order system, bifurcation curves, and Lyapunov exponents in the literature. Then, backstepping and sliding mode controller methods were applied to control the proposed system. It has been shown that these applied controllers can control the system even when the fractional degree of the system changes. However, in the backstepping method, it has been observed that the system approaches the desired equilibrium points for control different fractional-order systems. In addition, it has been observed that it can control both backstepping and sliding mode controller system when the fraction grades are high. However, due to the designed surface or its parameters, it was seen that the sliding mode controller came to the equilibrium point later. In addition, when the fractional degree is decreased, when the sliding mode control method is applied, it has been observed that the system reaches a different equilibrium point due to the dynamics of the system and the designed surface. In this direction, when the fractional degree decreases, it is concluded that different surface designs should be made from the literature, taking into account the different equilibrium points of the fractional chaotic system and the dynamics of the chaotic system. Nevertheless, both controllers can successfully regulate the system by obtaining control laws, even if the system is out of chaotic situations for specified fractional-order. In future studies, these methods will be used in the synchronization of different fractional-order systems, as well as fuzzy sliding mode, fuzzy backstepping control, adaptive sliding mode, adapting backstepping control techniques, which can be applied to systems with different fractional degrees.

References

1. Lorenz, E.N.: Deterministic nonperiodic flow. In: Universality in chaos, 2nd edn, pp. 367–378 (1963)
2. Kassim, S., Hamiche, H., Djennoune, S., Bettayeb, M.: A novel secure image transmission scheme based on synchronization of fractional-order discrete-time hyperchaotic systems. Nonlinear Dyn. **88**(4), 2473–2489 (2017). https://doi.org/10.1007/s11071-017-3390-8
3. Pashaei, S., Badamchizadeh, M.: A new fractional-order sliding mode controller via a nonlinear disturbance observer for a class of dynamical systems with mismatched disturbances. ISA Trans. **63**, 39–48 (2016). https://doi.org/10.1016/J.ISATRA.2016.04.003
4. Xu, C., Sun, Y., Gao, J., Qiu, T., Zheng, Z., Guan, S.: Synchronization of phase oscillators with frequency-weighted coupling. Sci. Rep.**6**, 1–9 (2016). https://doi.org/10.1038/srep21926

5. Singh, A.K., Yadav, V.K., Das, S.: Synchronization between fractional-order complex chaotic systems with uncertainty. Optik (Stuttg) **133**, 98–107 (2017). https://doi.org/10.1016/J.IJLEO.2017.01.017
6. Tavazoei, M.S., Haeri, M., Jafari, S., Bolouki, S., Siami, M.: Some applications of fractional calculus in suppression of chaotic oscillations. IEEE Trans. Ind. Electron. **55**(11), 4094–4101 (2008). https://doi.org/10.1109/TIE.2008.925774
7. Vaidyanathan, S., et al.: A 5-D multi-stable hyperchaotic two-disk dynamo system with no equilibrium point: Circuit design, FPGA realization and applications to TRNGs and image encryption. IEEE Access **9**, 81352–81369 (2021). https://doi.org/10.1109/ACCESS.2021.3085483
8. Sambas, A., et al.: A 3-D multi-stable system with a peanut-shaped equilibrium curve: circuit design, FPGA realization, and an application to image encryption. IEEE Access **8**, 137116–137132 (2020). https://doi.org/10.1109/ACCESS.2020.3011724
9. Tsafack, N., et al.: A new chaotic map with dynamic analysis and encryption application in internet of health things. IEEE Access **8**, 137731–137744 (2020). https://doi.org/10.1109/ACCESS.2020.3010794
10. Tsafack, N., Kengne, J., Abd-El-Atty, B., Iliyasu, A.M., Hirota, K., Abd EL-Latif, A.A.: Design and implementation of a simple dynamical 4-D chaotic circuit with applications in image encryption. Inf. Sci. (NY)**515**, 191–217 (2020). https://doi.org/10.1016/j.ins.2019.10.070
11. Tsafack, N., et al.: A memristive RLC oscillator dynamics applied to image encryption. J. Inf. Secur. Appl.**61**, 102944 (2021). https://doi.org/10.1016/j.jisa.2021.102944
12. Akgül, A., Rajagopal, K., Durdu, A., Pala, M.A., Boyraz, Ö.F., Yildiz, M.Z.: A simple fractional-order chaotic system based on memristor and memcapacitor and its synchronization application. Chaos Solitons Fractals **152**, 111306 (2021). https://doi.org/10.1016/j.chaos.2021.111306
13. Ni, J., Liu, L., Liu, C., Hu, X.: Fractional-order fixed-time nonsingular terminal sliding mode synchronization and control of fractional-order chaotic systems. Nonlinear Dyn. **89**(3), 2065–2083 (2017). https://doi.org/10.1007/S11071-017-3570-6
14. Pahnehkolaei, S.M.A., Alfi, A., Machado, J.A.T.: Fuzzy logic embedding of fractional-order sliding mode and state feedback controllers for synchronization of uncertain fractional chaotic systems. Comput. Appl. Math. **39**(3), 1–16 (2020). https://doi.org/10.1007/s40314-020-01206-7
15. Laarem, G.: A new 4-D hyper chaotic system generated from the 3-D Rösslor chaotic system, dynamical analysis, chaos stabilization via an optimized linear feedback control, it's fractional-order model and chaos synchronization using optimized fractional-order sliding mode control. Chaos Solitons Fractals **X**, 100063 (2021). https://doi.org/10.1016/J.CSFX.2021.100063
16. Mirrezapour, S.Z., Zare, A.: A new fractional sliding mode controller based on nonlinear fractional-order proportional integral derivative controller structure to synchronize fractional-order chaotic systems with uncertainty and disturbances. JVC/J. Vib. Control (2021). https://doi.org/10.1177/1077546320982453
17. Soukkou, A., Boukabou, A., Leulmi, S.: Prediction-based feedback control and synchronization algorithm of fractional-order chaotic systems. Nonlinear Dyn. **85**(4), 2183–2206 (2016). https://doi.org/10.1007/S11071-016-2823-0
18. Asemani, M.H., Majd, V.J.: Stability of output-feedback DPDC-based fuzzy synchronization of chaotic systems via LMI. Chaos Solitons Fractals **42**(2), 1126–1135 (2009). https://doi.org/10.1016/J.CHAOS.2009.03.012
19. Mofid, O., Mobayen, S., Khooban, M.H.: Sliding mode disturbance observer control based on adaptive synchronization in a class of fractional-order chaotic systems. Wiley Online Libr.**33**(3), 462–474 (2019). https://doi.org/10.1002/acs.2965
20. Jiang, C., Zada, A., Şenel, M.T., Li, T.: Synchronization of bidirectional N-coupled fractional-order chaotic systems with ring connection based on antisymmetric structure. Adv. Differ. Equ. **2019**(1), 1–16 (2019). https://doi.org/10.1186/s13662-019-2380-1
21. Danca, M.F.: Hidden chaotic attractors in fractional-order systems. Nonlinear Dyn. **89**(1), 577–586 (2017). https://doi.org/10.1007/S11071-017-3472-7

22. Benettin, G., Galgani, L., Giorgilli, A., Strelcyn, J.M.: Lyapunov characteristic exponents for smooth dynamical systems and for hamiltonian systems; a method for computing all of them. Part 1: theory. Meccanica **15**(1), 9–20 (1980). https://doi.org/10.1007/BF02128236
23. Balcerzak, M., Pikunov, D., Dabrowski, A.: The fastest, simplified method of Lyapunov exponents spectrum estimation for continuous-time dynamical systems. Nonlinear Dyn. **94**(4), 3053–3065 (2018). https://doi.org/10.1007/s11071-018-4544-z
24. Dabrowski, A.: Estimation of the largest Lyapunov exponent from the perturbation vector and its derivative dot product. Nonlinear Dyn. **67**(1), 283–291 (2012). https://doi.org/10.1007/s11 071-011-9977-6
25. Balcerzak, M., Pikunov, D.: The fastest, simplified method of estimation of the largest Lyapunov exponent for continuous dynamical systems with time delay. Mech. Mech. Eng. **21**(4), 985–994 (2017)
26. Danca, M.F., Kuznetsov, N.: Matlab code for Lyapunov exponents of fractional-order systems. Int. J. Bifurc. Chaos **28**(5), 1–14 (2018). https://doi.org/10.1142/S0218127418500670
27. Aguila-Camacho, N., Duarte-Mermoud, M.A., Gallegos, J.A.: Lyapunov functions for fractional-order systems. Commun. Nonlinear Sci. Numer. Simul. **19**(9), 2951–2957 (2014). https://doi.org/10.1016/j.cnsns.2014.01.022
28. Podlubny, I.: Fractional differential equations: an introduction to fractional derivatives, fractional differential equations, to methods of their solution and some of their applications. Elsevier (1998)
29. Gatto, A.E.: Product rule and chain rule estimates for fractional derivatives on spaces that satisfy the doubling condition. J. Funct. Anal. **188**(1), 27–37 (2002). https://doi.org/10.1006/jfan.2001.3836
30. Fujiwara, K., Georgiev, V., Ozawa, T.: Higher order fractional Leibniz rule. J. Fourier Anal. Appl. **24**(3), 650–665 (2018). https://doi.org/10.1007/S00041-017-9541-Y
31. Nyamoradi, N., Javidi, M.: Sliding mode control of uncertain unified chaotic fractional-order new lorenz-like system. Dyn. Contin. Discret. Impuls. Syst. Ser. B Appl. Algorithms **20**(1), 63–82 (2013)
32. Yuan, J., Shi, B., Zeng, X., Ji, W., Pan, T.: Sliding mode control of the fractional-order unified chaotic system. Abstr. Appl. Anal.**2013** (2013). https://doi.org/10.1155/2013/397504

Quantum Oscillations: A Promising Field for Secure Communication

Tanmoy Banerjee and Biswabibek Bandyopadhyay

Abstract Quantum communication has been identified as the most secured technique of future communication systems. Its security is directly tied up to the fundamental laws of quantum mechanics that never fail. The success of quantum communication hinges on the notion of oscillations and synchronization in the quantum regime. In this chapter, we discuss self-sustained oscillations and their synchronizations in the quantum regime. Unlike classical oscillations, quantum oscillations are bounded by several constraints of quantum mechanics. For example, although a prominent phase trajectory is possible in classical systems, quantum phase trajectories are broad due to the Heisenberg uncertainty principle. Here we systematically introduce the concept of quantum limit cycle using the open quantum formulation. Solutions of the quantum master equation and visualizations through the Wigner function are used to demonstrate the quantum limit cycle. Next, we discuss the possibility of synchronization in the quantum regime. Finally, we discuss some possible applications of quantum oscillations and synchronization in quantum key distribution and cybersecurity.

1 Introduction

Understanding of oscillation principles has been a central topic of interest in diverse fields from physics, chemistry, biology, engineering, and social science [1–3]. Oscillators are everywhere. We have hundreds of billions of neuronal oscillators in our brain that organize our cognitive system. In physics, Josephson junction and lasers are oscillators of quantum mechanical origin. Electronic communication systems

T. Banerjee (✉) · B. Bandyopadhyay
Chaos and Complex Systems Research Laboratory, Department of Physics, University of Burdwan, Burdwan 713 104, India
e-mail: tbanerjee@phys.buruniv.ac.in

© The Author(s), under exclusive license to Springer Nature Switzerland AG 2022
A. A. Abd El-Latif and C. Volos (eds.), *Cybersecurity*, Studies in Big Data 102,
https://doi.org/10.1007/978-3-030-92166-8_4

make extensive use of electronic oscillators. In ecology, population of species oscillates. All these oscillators have different time scales ranging from nano-second to few years. Although, the manifestation of oscillations are different in different oscillators all are governed by a common rule of nonlinear dynamics. Irrespective of its origin and manifestation, natural oscillators are essentially nonlinear [1, 4].

Perhaps the most celebrated model of nonlinear oscillator is the van der Pol (vdP) model [1, 5]. Starting its journey from the triode valve, it covers almost every branch of science. From an application point of view, engineering systems, such as electronic communication and information technology, widely exploit oscillations and synchronization principle for over a century.

With the recent advancement of quantum information processing and communications, understanding of oscillations in the quantum regime has gained much attention in recent years. However, the extension of the techniques used in the so called classical nonlinear dynamics to the quantum regime is not always straightforward. Understanding of nonlinear behavior in the quantum domain is based on the formalism of open quantum system that requires the solution of quantum master equations [6]. Also, phase space representation of quantum system which involves quasi probability function, e.g. Wigner function and Husimi function plays a crucial role [7]. Apart from the technique, the well known results of classical domain deviate in the quantum regime. This regime manifests several counterintuitive results that are not possible in the classical system. Therefore, to harness the richness of the quantum world in communication and computation one needs to understand the basic principles of quantum oscillations and synchronization.

In this chapter we will discuss self sustained oscillations in the quantum regime. We start with a classical van der Pol model and systematically establish its quantum analogue. Through numerical computation we visualize quantum limit cycle oscillations in the phase space. We discuss the basic difference between a classical and a quantum limit cycle. Then we will extend our study to understand the synchronization phenomenon in the quantum world. Finally, we discuss some possible applications of quantum synchronization in secure quantum communication. We believe that this work will help researchers to deepen their knowledge of fundamental oscillation principle in the quantum domain.

The chapter is organized in the following manner. The next section describes a classical van der Pol oscillator and a detailed derivation of its amplitude equation. Section 3 describes the quantum version of a van der Pol oscillator. It also demonstrates the quantum limit cycles and their time evolution. Section 4 presents some basic results of synchronization in the quantum regime. Finally, we discuss the relevance of the study in quantum information processing and secure communications.

2 Classical van der Pol oscillator

A van der Pol [5] oscillator has the following mathematical form:

$$\ddot{x} = -\omega^2 x + k_1 \dot{x} - 8k_2 x^2 \dot{x}, \tag{1}$$

where ω is the intrinsic frequency, k_1 is the gain factor corresponding to the linear pumping and k_2 is the loss factor corresponding to the nonlinear damping. Equation 1 can be written as a set of two first order differential equations as follows.

$$\dot{x} = \omega y, \tag{2a}$$

$$\dot{y} = -\omega x + (k_1 - 8k_2 x^2) y. \tag{2b}$$

Equation (2) shows a selfsustained oscillation (limit cycle) in phase space, which is shown in Fig. 1a.

We can express Eqs. (1) and (2) in terms of a complex amplitude $\alpha = x + iy$. Now we can follow the following steps:

$$
\begin{aligned}
\dot{\alpha} &= \dot{x} + i\dot{y} \\
&= \omega y - i\omega x + ik_1 y - 8ik_2 x^2 y \\
&= -i\omega(x + iy) + ik_1 \left(\frac{\alpha - \alpha^*}{2i} \right) \\
&\quad - 8ik_2 \left(\frac{\alpha + \alpha^*}{2} \right)^2 \left(\frac{\alpha - \alpha^*}{2i} \right) \\
&= -i\omega\alpha + \frac{k_1}{2}(\alpha - \alpha^*) - k_2(\alpha + \alpha^*)^2(\alpha - \alpha^*).
\end{aligned}
$$

Putting $\alpha = \eta e^{-i\phi}$ in both side we get,

$$
\begin{aligned}
e^{-i\phi}(\dot{\eta} - i\eta\dot{\phi}) &= -i\omega\eta e^{-i\phi} + \frac{k_1}{2}(\eta e^{-i\phi} - \eta e^{i\phi}) \\
&\quad - k_2(\eta e^{-i\phi} + \eta e^{i\phi})^2(\eta e^{-i\phi} - \eta e^{i\phi}) \\
\text{or,} \dot{\eta} - i\eta\dot{\phi} &= -i\omega\eta + \frac{k_1}{2}\eta(1 - e^{2i\phi}) \\
&\quad - k_2\eta^3(e^{-i\phi} + e^{i\phi})^2(1 - e^{2i\phi}) \\
&= -i\omega\eta + \frac{k_1}{2}\eta(1 - \eta e^{2i\phi}) \\
&\quad - k_2\eta^3(e^{-2i\phi} + e^{2i\phi} + 2)(1 - e^{2i\phi}).
\end{aligned}
\tag{3}
$$

Now at this point we apply the method of averaging. It can be done either by neglecting the terms containing the fast oscillations ($e^{\pm i\phi}, e^{\pm 2i\phi}, ...$) in the above equation

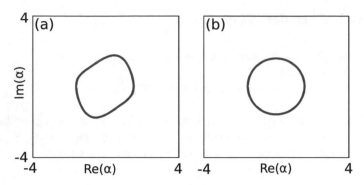

Fig. 1 Limit cycle in phase space governed by **a** Original van der Pol oscillator given by Eq. (2) and **b** Amplitude Eq. (4). Parameters are $k_1 = 1$, $k_2 = 0.2$, and $\omega = 2$

or by directly averaging the equations of $\dot{\eta}$ and $\dot{\phi}$ over one time period $T = \frac{2\pi}{\omega}$. Here, for simplicity we use the former technique. Neglecting the fast oscillation terms $(e^{\pm i\phi}, e^{\pm 2i\phi}, ...)$ in Eq. (3) we get,

$$\dot{\eta} - i\eta\dot{\phi} = -i\omega\eta + \frac{k_1}{2}\eta - k_2\eta^3$$

$$\text{or,} \quad e^{-i\phi}(\dot{\eta} - i\eta\dot{\phi}) = (-i\omega + \frac{k_1}{2} - k_2\eta^2)\eta e^{-i\phi}.$$

As $\eta = |\alpha|$, we can write the above equation as,

$$\dot{\alpha} = -i\omega\alpha + (\frac{k_1}{2} - k_2|\alpha|^2)\alpha \tag{4}$$

Equation 4 also shows a limit cycle oscillation of amplitude $\sqrt{\frac{k_1}{2k_2}}$, which is shown in Fig. 1b. Note that, Eq. (4) is valid in the weakly nonlinear (or quasi linear) condition only. It is also equivalent to the paradigmatic Stuart-Landau oscillator.

3 Quantum van der Pol Oscillator

Now the question is how to represent the quantum version of a van der Pol oscillator given by Eq. 4. Since a vdP oscillator is essentially a dissipative system, therefore it can not be represented by the unitary operations essentially used in the frame work of closed quantum mechanics. This is the reason we have to resort to the "open" quantum system where the operations of the system are governed by nonunitary rules and the notion of wave function is no longer useful [6]. In the open quantum system a harmonic oscillator is represented by the following dynamical equation in density matrix ρ:

$$\dot{\rho} = -i[\omega a^\dagger a, \rho], \tag{5}$$

where a and $a^\dagger$ are the bosonic annihilation and creation operators, respectively. Here and throughout the chapter we take $\hbar = 1$ as it does not affect the dynamics.

In order to get a "quantum" van der Pol oscillator our starting point is Eq. 5 and then introduce linear pumping and nonlinear damping in it systematically. In the formalism of open quantum system both can be done by considering that the harmonic oscillator (HO) is in the contact of an environment (or heat bath). The HO can exchange energy with the environment. The Hamiltonian of the system plus environment ($H_t = H + H_{env}$) still obeys the unitary operation as given in (5), however, the system hamiltonian H now obeys non-unitary operation, which is represented by the following quantum master equation (after Markove and Born approximation; for a detailed discussion see [7]):

$$\dot{\rho} = -i[\omega a^\dagger a, \rho] + \mathcal{L}(\rho), \tag{6}$$

where $\mathcal{L}(\rho)$ represents the Liouvilian operator that contains terms governing the interaction with the environment. This particular form of master equation is called the Lindblad master equation, which is widely used in the study of open quantum system. In the case of a quantum van der Pol oscillator the quantum master equation reads [8, 9]:

$$\dot{\rho} = -i[\omega a^\dagger a, \rho] + k_1 \mathcal{D}[a^\dagger](\rho) + k_2 \mathcal{D}[a^2](\rho), \tag{7}$$

where $\mathcal{D}[\hat{L}](\rho)$ is the Lindblad dissipator having the form $\mathcal{D}[L](\rho) = \hat{L}\rho\hat{L}^\dagger - \frac{1}{2}\{\hat{L}^\dagger\hat{L}, \rho\}$, where $\hat{L}$ is an operator. The implications of k_1 and k_2 in the quantum master equation is the following: k_1 controls the rate of single photon creation (equivalent to the linear pumping), and k_2 controls the same of two-photon absorption (equivalent to the nonlinear damping of classical case).

In the semiclassical limit, linear pumping dominates over the nonlinear damping (i.e., $k_1 > k_2$) and one approximates $\langle a \rangle \equiv \alpha$, and starting from the master Eq. (7) one arrives at the classical amplitude Eq. (4) by the following relation:

$$\langle \dot{a} \rangle = \text{Tr}(\dot{\rho} a)$$

3.1 Correspondence Between Master Equation and Amplitude Equation

In the above relation we have used the following technique: In quantum optics we know the clear correspondence between the average of annihilation operator ($\langle a \rangle$) and the complex amplitude (α), they are equivalent ($\langle a \rangle \equiv \alpha$) [10]. This property can bridge the quantum master equation and the amplitude equation. Let us consider the simplest form of the master equation as $\dot{\rho} = -i[H, \rho] + \mathcal{D}[L](\rho)$. Now the average of any operator '$\hat{O}$' is given by $\left\langle \hat{O} \right\rangle = Tr(\rho \hat{O})$. So the dynamical equation of $\left\langle \hat{O} \right\rangle$ is given by,

$$
\begin{aligned}
\frac{d\left\langle \hat{O} \right\rangle}{dt} &= \frac{d}{dt} Tr(\rho \hat{O}) \\
&= Tr(\hat{O}\frac{d\rho}{dt}) \\
&= i \left\langle [H, \hat{O}] \right\rangle + Tr(\hat{O}\mathcal{D}[L](\rho)) \\
&= i \left\langle [H, \hat{O}] \right\rangle + \left\langle \tilde{\mathcal{D}}[L](\hat{O}) \right\rangle,
\end{aligned}
\tag{8}
$$

where $\tilde{\mathcal{D}}[L](\hat{O})$ is called the 'Adjoint operator', having the following form:

$$
\begin{aligned}
\tilde{\mathcal{D}}[L](\hat{O}) &= L^\dagger \hat{O} L - \frac{1}{2}\{L^\dagger L, \hat{O}\} \\
&= \frac{1}{2}\left(L^\dagger[\hat{O}, L] + [L^\dagger, \hat{O}]L\right)
\end{aligned}
\tag{9}
$$

Therefore, following the above process shown in Eq. (8) and using the master Eq. (7), we want to evaluate the dynamical equation of expectation value of the annihilation operator ($\langle a \rangle$).

$$
\begin{aligned}
\langle \dot{a} \rangle &= i \left\langle [\omega a^\dagger a, a] \right\rangle + k_1 \left\langle \tilde{\mathcal{D}}[a^\dagger](a) \right\rangle \\
&\quad + k_2 \left\langle \tilde{\mathcal{D}}[a^2](a) \right\rangle \\
&= i\omega \left\langle [a^\dagger a, a] \right\rangle + \frac{k_1}{2}\left\langle (a[a, a^\dagger] + [a, a]a^\dagger) \right\rangle \\
&\quad + \frac{k_2}{2}\left\langle (a^{\dagger 2}[a, a^2] + [a^{\dagger 2}, a]a^2) \right\rangle \\
&= -i\omega \langle a \rangle + \frac{k_1}{2}\langle a \rangle - k_2 \left\langle a^\dagger a^2 \right\rangle,
\end{aligned}
\tag{10}
$$

which is the same as Eq. (4) as $\alpha \equiv \langle a \rangle$. Similarly, we can derive the dynamical equation of expectation value of creation operator ($\langle a^\dagger \rangle$), which has the following form:

$$
\langle \dot{a^\dagger} \rangle = i\omega \left\langle a^\dagger \right\rangle + \frac{k_1}{2}\left\langle a^\dagger \right\rangle - k_2 \left\langle a^{\dagger 2} a \right\rangle
\tag{11}
$$

This equation is equivalent to the complex conjugate form of Eq. (4).

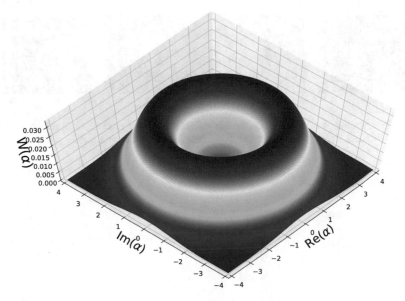

Fig. 2 A three dimensional view of the quantum limit cycle. Parameters are $\omega = 2$, $k_1 = 1$ and $k_2 = 0.2$

3.2 Visualising Quantum Limit Cycle

A numerical solution of (7) gives the time evolution of ρ. But how to visualize the limit cycle? The most useful way is to observe Wigner function of the system in the phase space. The Wigner function is a quasi-probability function that gives the probability of finding the system around a point in the phase space [7, 11]. Unlike classical probability, it can be negative (that is why the term quasi-probability). A negative Wigner function is the indicator of a strict quantum process that has no counterpart in classical domain. A formal definition of the Wigner function is

$$W(x, p) = \frac{1}{2\pi\hbar} \int\limits_{\infty}^{-\infty} dy\, e^{-ipy/\hbar} \langle x + \frac{y}{2}|\rho|x - \frac{y}{2}\rangle. \tag{12}$$

Note the essential difference between the classical and quantum phase space. In a classical phase space, phase point is distinct and therefore, the phase trajectories are prominent. However, due to the Heisenberg uncertainty principle, in the quantum phase space phase points and phase trajectories are smeared blobs and regions, respectively.

Using QuTip [12] module in Python we numerically solve Eq. 7. Figure 2 shows the Wigner function of the quantum limit cycle for the following parameters: $k_1 = 1$ and $k_2 = 0.2$ (without any loss of generality, throughout the paper we take $\omega = 2$). It can be clearly seen that the maximum probability occurs in a ring-shaped region

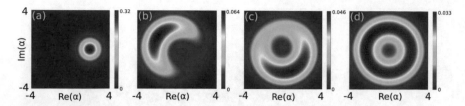

Fig. 3 Time evolution of the limit cycle: **a** Initial condition is a coherent state on the x-axis. **b** $t = 5$. **c** $t = 10$. **d** Steady state limit cycle. Parameters are $\omega = 2$, $k_1 = 1$ and $k_2 = 0.2$

Fig. 4 Distribution of the occupation of Fock levels of Fig. 3d

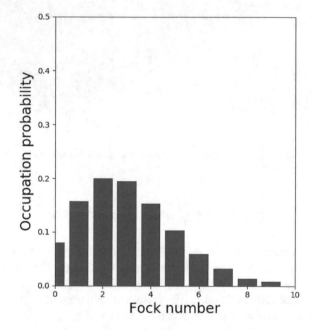

of the phase space. this is nothing but the quantum limit cycle of the quantum vdP oscillator. The probability sharply falls towards the origin and outside the ring.

Next, we try to understand the time evolution of the limit cycle. For this we consider a coherent state as the initial condition. At $t = 0$ it is shown in Fig. 3a. As time passes by, the blob rotates around the origin and at the same time it gets dispersed. Figure 3b, c show these intermediate states at $t = 5$ and $t = 10$, respectively. At the steady state the probability function resides on a blurred ring shaped region, which is the quantum limit cycle. Figure 3d demonstrates it, which is the 2D view of Fig. 2. Note another basic difference between a classical and quantum limit cycle. In the classical case, a phase point rotates with time (with its nominal frequency). However, in the steady state, a quantum limit cycle is "present" along the ring uniformly. In that sense a quantum limit cycle is indeed a steady state.

Finally, we demonstrate how number states (or known as Fock levels) are populated during a limit cycle oscillations. Figure 4 presents the occupation probability of

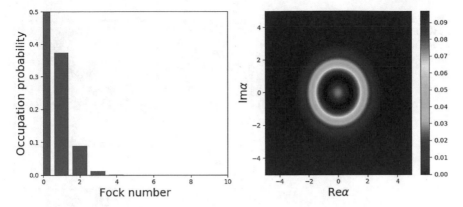

Fig. 5 Quantum limit cycle in the deep quantum regime. Left panel: Distribution of the occupation of Fock levels. Right panel: Wigner function. Parameters are $\omega = 2$, $k_1 = 1$ and $k_2 = 3$

the number states. It shows a Poisson like distribution. An inspection of Fock levels is useful specially in the deep quantum regime. In this regime nonlinearity becomes strong, i.e., dissipation dominates over the pumping. This condition is achieved by setting $k_2 > k_1$. The Fock distribution and Wigner function of a quantum limit cycle in the deep quantum regime ($k_1 = 1$ and $k_2 = 3$) is shown in Fig. 5. Note that in the deep quantum regime photons predominantly populate the ground state and only a few higher Fock levels are populated. Also, the ring shape of the Wigner function is quenched. One interesting fact is that, even in the limit of $k_2 \to \infty$, i.e., infinite damping, a quantum limit cycle survives. This is because of the fact that the two-photon absorption process can not entirely relax to the $|0\rangle$ state. In the limiting case of $k_2 \to \infty$, the steady state density matrix is [8]: $\rho_{ss} = \frac{2}{3}|0\rangle\langle 0| + \frac{1}{3}|1\rangle\langle 1|$, i.e., the oscillation still persists unlike a classical vdP oscillator.

4 Quantum Synchronization

In order to use quantum vdP oscillators in applications like quantum communication, we must inspect about its ability to synchronize with external signals or with other oscillators [13, 14]. The topic of quantum synchronization is new and still in its infant state. Two seminal papers [8, 9] explored quantum synchronization in detail. Here we will only discuss the main results. It has been found that the inherent quantum noise resists a perfect frequency-locking in the quantum regime.

In order to understand quantum synchronization, let us discuss the synchronization of a quantum vdP oscillator to an external signal as reported in Ref. [9]. The master equation of quantum vdP oscillator with an external harmonic drive [9] can be given by,

$$\dot{\rho} = -i[\hat{H}, \rho] + k_1 \mathcal{D}[a^\dagger](\rho) + k_2 \mathcal{D}[a^2](\rho), \tag{13}$$

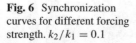

Fig. 6 Synchronization curves for different forcing strength. $k_2/k_1 = 0.1$

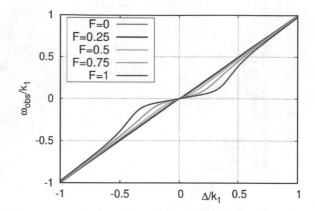

where $\hat{H} = \Delta a^\dagger a + iF(a - a^\dagger)$. Here $\Delta = \omega - \omega_d$ is the detuning frequency, with ω as the frequency of the oscillator and ω_d as the frequency of the drive with drive strength F.

The most interesting scenario of synchronization can be seen as the frequency locking of a self-sustained oscillator under an externally driven harmonic force. There may be a finite range of detuning for which oscillator remains in a frequency locked state with the external drive. In a frequency power spectrum, at which frequency the power is maximum is called the observed frequency (ω_{obs}). For lower frequency entrainment, ω_{obs} attains a value near the detuning frequency. But as we move towadrs strong entrainment zone by suitably varying the drive strength, ω_{obs} shifts towards zero, indicating better frequency entrainment ($\omega \approx \omega_d$).

Figure 6 (at $k_2/k_1 = 0.1$) shows different variations of ω_{obs}/k_1 with the detuning Δ/k_1 for different driving strength. It is clear that for lower values of the driving strength, ω_{obs}/k_1 remains approximately equal to Δ/k_1 even at a very low value of detuning. But upon increasing the driving strength, for lower values of Δ/k_1, ω_{obs}/k_1 tends to shift towards zero, which indicates a stronger frequency entrainment.

In the above case, although driving helps the shifting of the peak of power spectrum towards $\omega_{obs}/k_1 = 0$, for any value of Δ/k_1 the observed frequency does not become exactly zero. That is why we can say that complete synchronization does not occur by applying only the external harmonic drive. This happens because of the presence of inherent quantum noise in the system. This quantum noise hinders the synchronization and external drive is not enough to overcome this hindrance. However, research is going on to improve the quality of quantum synchronization [15].

5   Discussion: Applications in Secure Communication Systems

In this chapter we have discussed a systematic way to understand quantum limit cycle oscillations. Starting from a classical van der Pol oscillator we have arrived at the quantum master equation that gives quantum limit cycle. We discussed both the cases of weak and deep quantum regime. Further, we discussed the possibility of synchronization in quantum self sustained oscillators. This primer provides a first step in understanding quantum oscillation that has been rapidly growing as an alternative of chaos in secure communication systems.

Several experimental techniques have been proposed to implement quantum vdP oscillators. A few feasible set up are ion-trap experiment [8], optomechanical set up [9], and superconducting circuits [16]. Therefore, it is expected to realize quantum self sustained oscillators in the real system in a controlled set up.

Quantum oscillation and their synchronization open up a myriads of possibilities in the application of quantum communication, cryptography, and quantum computation [17, 18]. Here we can harness the richness of the quantum world (e.g., no cloning principle) in order to make things more secure and fast. Specifically, the potential of quantum synchronization in Quantum key distribution (QKD) has already been realized in theory and implemented in experiment [19]. The QKD, originally introduced in 1984 as a communication protocol for the generation and share of a secret key, has been identified as a promising technology for the secure future communication networks [19]. The security of the QKD is linked to the fundamental laws of quantum mechanics, which never break down. This shows the strength of the protocol in comparison with its classical counterpart. Several recent experiments verify the practical implementation of the QKD protocol, e.g., in telecommunication through fibers, the daylight free-space channel communication and the satellite-to-ground channel communication [19, 20]. Moreover, in an another context, the recent discovery of symmetry-breaking squeezed steady states [21–23] in the quantum oscillators opens up the possibility of their applications in spectroscopy and the detection of gravitational wave [24–26]. Further, with the recent advancement of quantum information processing and communications, quantum walks have gained much attention in recent years [27–31]. Exploring the connection between the quantum dissipative dynamics and the quantum walk is a promising topic of research. Therefore, it can be said that the quantum oscillation and synchronization is a promising field that is waiting for its application potentiality to be fully explored.

Several open challenges remain in order to fully understand quantum synchronization, and therefore, to apply those ideas in practical applications. A few topics that need immediate attentions are: (i) Understanding of quantum oscillation in the strong nonlinear regime. Recent advancements in this topic (e.g., obtaining relaxation oscillations [32, 33]) guides the path to quantize a classical nonlinear system described by differential equations. Quantization of chaotic classical oscillators in order to get their equivalent master equation is an open challenge for the researchers. (ii) Understanding of network dynamics of quantum oscillators under arbitrary coupling topology is also an interesting (and difficult) topic of research, as that may unravel the quantum equivalent of partial synchronization states [34–37]. (iii) Finding connections between synchronization and the paradigmatic quantum phenomenon, namely entanglement [38, 39] is an open challenge for researchers; a complete understanding of their relation would lead to much more secure futuristic communication system, such as quantum teleportation.

Acknowledgements B.B. acknowledges the financial assistance from the University Grants Commission (UGC), India. T. B. acknowledges the financial support from the Science and Engineering Research Board (SERB), Govt. of India, in the form of a Core Research Grant [CRG/2019/002632].

References

1. Pikovsky, A., Rosenblum, M., Kurths, J.: Synchronization: a Universal Concept in Nonlinear Sciences. Cambridge University Press, England (2003)
2. Strogatz, S.: Sync: the Emerging Science of Spontaneous Order. Penguin, UK (2004)
3. Biswas, D., Banerjee, T.: Time-Delayed Chaotic Dynamical Systems. Springer-Nature, Cham (2018)
4. Banerjee, T., Biswas, D.: Nonlinear Dyn. **73**, 2025–2048 (2013)
5. van der Pol, B.: Philos. Mag. **43**, 700 (1922)
6. Breuer, H.P., Petruccione, F.: The Theory of Open Quantum Systems. Oxford University Press, Oxford, UK (2002)
7. Carmichael, H.J.: Statistical Methods in Quantum Optics 1. Springer (1999)
8. Lee, T.E., Sadeghpour, H.R.: Phys. Rev. Lett. **111**, 234101 (2013)
9. Walter, S., Nunnenkamp, A., Bruder, C.: Phys. Rev. Lett. **112**, 094102 (2014)
10. Gerry, C., Knight, P.: Introductory Quantum Optics. Cambridge University Press, Cambridge, England (2005)
11. Weinbub, J., Ferry, D.K.: Appl. Phys. Rev. **5**, 041104 (2018)
12. Johansson, J., Nation, P., Nori, F.: Comput. Phys. Commun. **184**, 1234 (2013)
13. Morgan, L., Hinrichsen, H.: J. Stat. Mech. **28**, P09009 (2015)
14. Walter, S., Nunnenkamp, A., Bruder, C.: Ann. der Phys. **527**, 131 (2015)
15. Sonar, S., Hajdušek, M., Mukherjee, M., Fazio, R., Vedral, V., Vinjanampathy, S., Kwek, L.C.: Phys. Rev. Lett. **120**, 163601 (2018)
16. Lörch, N., Nigg, S.E., Nunnenkamp, A., Tiwari, R.P., Bruder, C.: Phys. Rev. Lett. **118**, 243602 (2017)
17. Nielsen, M.A., Chuang, I.L.: Quantum Computation and Quantum Information. Cambridge University Press, UK (2010)
18. Abd El-Latif, A.A., Abd-El-Atty, B., Venegas-Andraca, S.E., Mazurczyk, W.: Fut. Gener. Comput. Syst. **100**, 893–906 (2019)
19. Calderaro, L., Stanco, A., Agnesi, C., Avesani, M., Dequal, D., Villoresi, P., Vallone, G.: Phys. Rev. Applied **13**, 054041 (2020)
20. Liao, S.K., Yong, H.L., Liu, C.E.A.: Nat. Photon **11**, 509 (2017)
21. Bandyopadhyay, B., Khatun, T., Biswas, D., Banerjee, T.: Phys. Rev. E **102**, 062205 (2020)
22. Bandyopadhyay, B., Banerjee, T.: Chaos **31**, 063109 (2021)
23. Bandyopadhyay, B., Khatun, T., Banerjee, T.: Phys. Rev. E **104**, 024214 (2021)
24. Iles-Smith, J., Nazir, A., McCutcheon, D.P.: Nat. Comm. **10**, 3034 (2019)
25. Lloyd, S., Braunstein, S.L.: Phys. Rev. Lett. **82**, 1784 (1999)
26. Goda, K., Mikhailov, O.M.E.E., Saraf, S., Adhikari, R., McKenzie, K., Ward, R., Vass, S., Weinstein, A.J., Mavalvala, N.: Nat. Phys. **4**, 472 (2008)

27. Abd El-Latif, A.A., Abd-El-Atty, B., Mehmood, I., Muhammad, K., Venegas-Andraca, S.E., Peng, J.: Inf. Process. Manag. **54**(4), 102549 (2021)
28. Alanezi, A., Abd-El-Atty, B., Kolivand, H., Abd El-Latif, A.A.: 2021 1st International Conference on Artificial Intelligence and Data Analytics (CAIDA), pp. 176–181 (2021)
29. Abd-El-Atty, B., Iliyasu, A.M., Alanezi, A., Abd El-Latif, A.A.: Opt. Lasers Eng. **138**, 106403 (2021)
30. Abd-El-Atty, B., Iliyasu, A.M., Alaskar, H., Abd El-Latif, A.A.: Sensors **20**(11), 3108 (2020)
31. Abd El-Latif, A.A., Abd-El-Atty, B., Mazurczyk, W., Fung, C., Venegas-Andraca, S.E.: IEEE Trans. Netw. Serv. Manag. **17**(1), 118–131 (2020)
32. Chia, A., Kwek, L.C., Noh, C.: Phys. Rev. E **102**, 042213 (2020)
33. Ben Arosh, L., Cross, M.C., Lifshitz, R.: Phys. Rev. Res. **3**, 013130 (2021)
34. Bastidas, V.M., Omelchenko, I., Zakharova, A., Schöll, E., Brandes, T.: Phys. Rev. E **92**, 062924 (2015)
35. Banerjee, T., Biswas, D., Ghosh, D., Schöll, E., Zakharova, A.: Chaos **28**, 113124 (2018)
36. Banerjee, T.: EPL (Europhys. Lett.) **110**, 60003 (2015)
37. Bandyopadhyay, B., Khatun, T., Dutta, P.S., Banerjee, T.: Chaos Solitons Fractals **139**, 110289 (2020)
38. Lee, T.E., Chan, C.K., Wang, S.: Phys. Rev. E **89**, 022913 (2014)
39. Cattaneo, M., Giorgi, G.L., Maniscalco, S., Paraoanu, G.S., Zambrini, R.: Annalen der Physik **533**(5), 2100038 (2021)

Synchronization of Chaotic Electroencephalography (EEG) Signals

Jessica Zaqueros-Martinez, Gustavo Rodriguez-Gomez,
Esteban Tlelo-Cuautle, and Felipe Orihuela-Espina

Abstract Synchronization of chaotic signals often considers a master-slave paradigm where a slave chaotic system is required to follow the master also chaotic. Most times in literature both systems are known, but synchronization to some unknown master has a potentially large range of applications, for example, EEG based authentication. We aim to test the feasibility of fuzzy control to systematically synchronize a chaotic EEG record. In this chapter, we study the suitability of two chaotic systems and the companion fuzzy control strategies under complete and projective synchronization to synchronize to EEG records. We used two public EEG datasets related to the genetic predisposition to alcoholism and with detecting emotions respectively. We present a comparative study among fuzzy control strategies for synchronization of chaotic systems to EEG records on selected datasets. As expected, we observed success and failures alike on the synchronization highlighting the difficulty in achieving this kind of synchronization, but we interpret this as advantageous for purposes of the suggested domain application. With successful synchronizations, we confirm that synchronization is feasible. With unsuccessful synchronizations, we illustrate that synchronization of chaotic systems does not follow a simple one-size-fits-all recipe and we attempt to gain insight for future research. The same chaotic system may succeed or fail depending on its companion type of synchronization and controller design.

J. Zaqueros-Martinez (✉) · G. Rodriguez-Gomez · E. Tlelo-Cuautle · F. Orihuela-Espina
INAOE, Puebla, Mexico
e-mail: jessizaqueros@inaoep.mx

G. Rodriguez-Gomez
e-mail: grodrig@inaoep.mx

E. Tlelo-Cuautle
e-mail: etlelo@inaoep.mx

F. Orihuela-Espina
e-mail: f.orihuela-espina@inaoep.mx

© The Author(s), under exclusive license to Springer Nature Switzerland AG 2022
A. A. Abd El-Latif and C. Volos (eds.), *Cybersecurity*, Studies in Big Data 102,
https://doi.org/10.1007/978-3-030-92166-8_5

83

1   Introduction

Chaos synchronization has emerged as an area of great interest because of the substantial theoretical challenge it represents but also because of the many potential practical applications. The latter include secure communications [19, 33], chemical reactions [10, 43], aerospace [10, 41], signal processing [15, 33], laser physics [10, 43], powers systems [41] and biological systems [37, 43], financial risk systems [28], among others.

Chaos synchronization refers to the problem of enforcing two chaotic dynamic systems to eventually evolve at the unison in all future events when departing from different initial conditions [26]. Chaos synchronization can be coarsely split into two major subproblems. Those in which one of the systems is enforced to follow another, known as master-slave, and those in which both systems are allowed to influence its counterpart, referred to as coupling. In the former case, the state of the slave system is guided by that of the master.

Regardless of the subproblem being dealt with, synchronization differs according to the choice of the error function. There exist a number of options; complete or identical synchronization, anti-synchronization, phase synchronization and projective synchronization, among others. Perhaps the most common is complete synchronization where the error is defined as the difference between the state variables of both systems. In the case of the master-slave configuration, that is the difference between the state variables of the slave and the master respectively [21, 22, 35, 38, 39]. A conceptually small modification sees that difference scaled by a factor different from zero, which corresponds to the projective synchronization [42].

Often in literature, it happens that the analytical expressions of both systems being synchronized are fully characterized; that is, their system of equations is known. The study of known systems gives us great insight into the problem of chaos synchronization. But for practical purposes, synchronizing chaotic systems where at least one, or both of the systems involved, are not analytically characterized is of great interest. Examples include financial applications such as the stock exchange, or climate applications such as weather forecasting. Finally, another application domain of particular interest to this chapter is biometric systems.

Biometric systems aim at univocally identifying a person (or actually any other biological organisms) from its biosignals [36] or behavioural characteristics [1]. These biosignals are traces of information, whether static or dynamic, sensed from the organism body, e.g. electrophysiological signals, hemodynamic readings, iris pictures, voice, facial expression, fingerprint, among others. Biometric systems have already exhibit usefulness in tasks of authentication with implications for security. Biometric traces include fingerprint, voice or facial recognition [31, 34]. Exemplary applications may include online banking for payment authorization and identity verification [31], governance for authentication services for citizens, access to government facilities or education for servicing students [1]. For a discussion in greater length, the reader is directed to the wealth of existing literature on the topic e.g. [36].

Electroencephalography-based biometry is often recognized as more confidential, sensitive, and hard to steal and replicate than other alternatives [7]. Hence, they hold great promise to provide a far more secure biometric approach for user identification and authentication. An electroencephalogram (EEG) non-invasively records the brain electrical activity through electrodes on the scalp [18]. It is well known that EEG recordings exhibit chaotic behaviour [9, 23]. However, despite the very sophisticated models from the field of computational neuroscience [5, 27, 32], there is thus far no explicit system of equations to precisely express a given EEG.

Putting all together, EEG-based authentication makes a potentially high impact application of synchronization of chaotic systems where at least one of the involved systems is not analytically known. The goal of this research is to test the feasibility of fuzzy controls to systematically synchronize a chaotic EEG record. In this chapter, we study the suitability of Chua and Rossler chaotic systems and the fuzzy controls strategies under the complete and projective synchronization for synchronizing EEG records. The differences in the retrieved system parameters are what open the door to biometric applications. In our comparative study, we show that the success of synchronizing a particular chaotic slave to a given EEG master depends on the choice of the type of error function as well as the design of the control. Employing fuzzy control facilitates switching the chaotic slave because the design of the control only depends on the gain matrices and the controlling parameter, which only require recalculation whenever the slave changes.

The novelty of this synchronization with respect to others in the literature on synchronization of chaotic systems is that the EEG has no known analytical model that generates it. We here worked with two datasets from arbitrary public repositories. This arbitrariness is intentional. We intended to emphasize the message that a particular controlled stimulation is not required.

Synchronization of chaotic slaves to the unknown EEG chaotic master can be used both for basic neuroscience, to better understand the EEG, as well as for clinical, as a biomarker of pathologies, and applied neuroscience, for instance for brain-computer interfaces, or biometrics.

2 Fuzzy Models of Chaotic Systems

The use of fuzzy theory has been said to permit a generalization of information by means of the introduction of imprecision [12]. Thus, fuzzy models can facilitate many analytical tasks over chaotic systems. They have proved to be particularly useful in control applications [12] where chaos synchronization is nothing but a specific case.

The fuzzy modelling of chaotic systems for the purposes of control involves their fuzzification and later defuzzification. There are several options in the literature to implement such operations, e.g. the fuzzy models of Mamdani [16] and Takagi-Sugeno (T–S) fuzzy models [29]. In this work, we opt for T–S fuzzy models which locally linearize a chaotic system. In the following sections, we present the fuzzification step using this model following the methodology described in [13, 14].

2.1 Chua System

Chua's two-diode system (referenced in this chapter using its author name for simplicity), has its origin on an electronic circuit with non-linear feedback [3]. This
system is described by the third-order autonomous differential equation in Eq. (1).

$$
\begin{aligned}
\dot{x}_1 &= \alpha(x_2 - x_1 - \psi(x_1)) \\
\dot{x}_2 &= \quad x_1 - x_2 + x_3 \\
\dot{x}_3 &= \quad -(\beta x_2 + \gamma x_3)
\end{aligned}
\tag{1}
$$

where

$$
\psi(x_1) = m_1 x_1 + \frac{1}{2}(m_0 - m_1)\mathrm{sat}(x_1)
$$

$$
\mathrm{sat}(x_1) = \begin{cases}
-1 & \text{if } x_1 \le -1 \\
x_1 & \text{if } |x_1| < 1 \\
1 & \text{if } x_1 \ge 1
\end{cases}
$$

and x_1, x_2, x_3 are state variables, α, β and γ are parameters [4]. Given a certain
parametrization, this system is known to exhibit a hidden attractor [4, 11]. The chaotic
behaviour of the system around the hidden attractor with parameters $\alpha = 8.4562$, $\beta =
12.0732$, $\gamma = 0.0052$, $m_0 = -0.1768$, $m_1 = -1.1468$, and initial condition $x_0 =
[-3.7727, -1.3511, 4.6657]^T$ [8] is illustrated in Fig. 1.

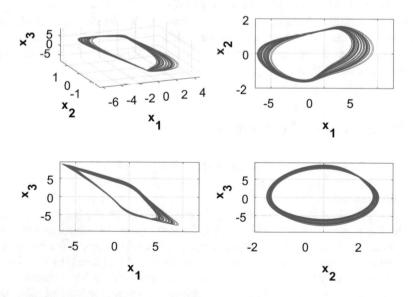

Fig. 1 A hidden attractor in Chua's chaotic system

Assume that $x_1 \in [-d, d]$ and $d > 0$, then, the fuzzy model which exactly represents the system in Eq. (1) is given by Eq. (2):

$$\text{Rule 1}: \quad \textbf{IF } x_1(t) \text{ is } F_1 \textbf{ THEN } \dot{x}(t) = A_1 x(t) + b_1$$
$$\text{Rule 2}: \quad \textbf{IF } x_1(t) \text{ is } F_2 \textbf{ THEN } \dot{x}(t) = A_2 x(t) + b_2$$

(2)

where $x = [x_1, x_2, x_3]^T$, $b_1 = b_2 = [0, 0, 0]^T$, $F_1(x_1) = \frac{1}{2}\left(1 - \frac{\phi(x_1)}{d}\right)$, $F_2(x_1) = 1 - F_1(x_1)$, and:

$$A_1 = \begin{pmatrix} \alpha(d-1) & \alpha & 0 \\ 1 & -1 & 1 \\ 0 & -\beta & -\gamma \end{pmatrix}$$

$$A_2 = \begin{pmatrix} -\alpha(d+1) & \alpha & 0 \\ 1 & -1 & 1 \\ 0 & -\beta & -\gamma \end{pmatrix}$$

$$\phi(x_1) = \begin{cases} \frac{\psi(x_1)}{x_1} & x_1 \neq 0 \\ m_0 & x_1 = 0 \end{cases}$$

The final output of the Chua's fuzzy system is inferred by Eq. (3):

$$\dot{x} = \sum_{i=1}^{2} F_i(x_1(t))\{A_i x(t) + b_i\}$$

(3)

2.2 Rossler System

Although we are not aware of a physical interpretation of Rossler's system, we know that historically, Rossler was looking for a chaotic system that could generate a single spiral [24]. His chaotic system is given by Eq. (4).

$$\dot{x}_1 = -x_2 - x_3$$
$$\dot{x}_2 = x_1 + ax_2$$
$$\dot{x}_3 = bx_1 - (c - x_1)x_3$$

(4)

where x_1, x_2, x_3 are state variables, and a, b and c are the system parameters [30]. The chaotic behaviour of the system with parameters $a = 0.34, b = 0.4, c = 4.5$ and initial conditions $x_0 = [1.5, 2.0, 2.0]^T$ [30] is illustrated in Figure 2.

Assume that $x_1 \in [c - d, c + d]$ and $d > 0$. Then, the fuzzy model which exactly represents the system in Eq. (4) is given by Eq. (5):

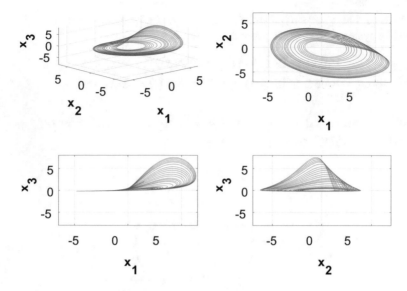

Fig. 2 Attractor in Rossler's chaotic system

$$\text{Rule 1}: \quad \textbf{IF } x_1(t) \text{ is } F_1 \textbf{ THEN } \dot{x}(t) = A_1 x(t) + b_1$$
$$\text{Rule 2}: \quad \textbf{IF } x_1(t) \text{ is } F_2 \textbf{ THEN } \dot{x}(t) = A_2 x(t) + b_2 \tag{5}$$

where $x = [x_1, x_2, x_3]^T$, $b_1 = b_2 = [0, 0, 0]^T$, $F_1(x_1) = \frac{1}{2}\left(1 + \frac{c-x_1}{d}\right)$, $F_2(x_1) = \frac{1}{2}\left(1 - \frac{c-x_1}{d}\right)$ and:

$$A_1 = \begin{pmatrix} 0 & -1 & -1 \\ 1 & a & 0 \\ b & 0 & -d \end{pmatrix}$$

$$A_2 = \begin{pmatrix} 0 & -1 & -1 \\ 1 & a & 0 \\ b & 0 & d \end{pmatrix}$$

Analogously to Chua's defuzzification, the final output of the Rossler's fuzzy system is inferred by Eq. (6):

$$\dot{x} = \sum_{i=1}^{2} F_i(x_1(t))\{A_i x(t) + b_i\} \tag{6}$$

3 Fuzzy Complete Synchronization

Let be a master system $\dot{x}(t) = f(x(t))$ where $x(t) = (x_1(t), x_2(t), \ldots, x_n(t)) \in \mathbb{R}^n$. This system can be fuzzily characterized as Eq. (7):

$$\text{Rule 1 : } \textbf{IF } x_1(t) \text{ is } F_1 \textbf{ THEN } \dot{x}(t) = A_1 x(t) + b_i$$
$$\text{Rule 2 : } \textbf{IF } x_1(t) \text{ is } F_2 \textbf{ THEN } \dot{x}(t) = A_2 x(t) + b_i \tag{7}$$

and the final output of the master fuzzy system is presented in Eq. (8).

$$\dot{x} = \sum_{i=1}^{2} F_i(x_1(t))\{A_i x(t) + b_i\} \tag{8}$$

A slave fuzzy system would be given by Eq. (9).

$$\text{Rule 1 : } \quad \textbf{IF } y_1(t) \text{ is } F_1 \quad \textbf{THEN } \dot{y}(t) = A_1 y(t) + b_1 + Bu(t)$$
$$\text{Rule 2 : } \quad \textbf{IF } y_1(t) \text{ is } F_2 \quad \textbf{THEN } \dot{y}(t) = A_2 y(t) + b_2 + Bu(t) \tag{9}$$

where $y(t) = (y_1(t), y_2(t), \ldots, y_n(t) \in \mathbb{R}^n$. B is an input matrix and $u(t)$ is the controller.

Slave defuzzification is given by Eq. (10).

$$\dot{y} = \sum_{i=1}^{2} F_i(y_1(t))\{A_i y(t) + b_i\} + Bu(t), \tag{10}$$

In complete synchronization, error is defined as $e(t) = y(t) - x(t)$. Taking the derivative of the error and substituting into Eqs. (10) and (8) we can derive Eq. (11).

$$\dot{e}(t) = \sum_{i=1}^{2} F_i(y_1(t))\{A_i y(t) + b_i\} - \sum_{i=1}^{2} F_i(x_1(t))\{A_i x(t) + b_i\} + Bu(t) \tag{11}$$

The fuzzy controller design, $u(t)$, is given in Eq. (12).

$$u(t) = -\sum_{i=1}^{2} F_i(y_1(t))\{C_i y(t) + b_i\} + \sum_{i=1}^{2} F_i(x_1(t))\{C_i x(t) + b_i\} \tag{12}$$

We seek $\|e(t)\| \to 0$ when $t \to \infty$. Substituting Eq. (12) in Eq. (11), one gets Eq. (13).

$$\dot{e}(t) = \sum_{i=1}^{2} F_i(y_1(t))\{(A_i - BC_i)y(t)\} - \sum_{i=1}^{2} F_i(x_1(t))\{(A_i - BC_i)x(t)\} \qquad (13)$$

The gain matrices C_i are calculated during the design process. With the idea of linearization given in [14, 30], if there exist gain matrices such that:

$$\{(A_1 - BC_1) - (A_2 - BC_2)\}^T \times \{(A_1 - BC_1) - (A_2 - BC_2)\} = 0 \qquad (14)$$

then, the total error of the system becomes linear when $\dot{e}(t) = Ge(t)$ according to the fuzzy controller in Eq. (12), where $G = A_1 - BC_1 = A_2 - BC_2$. Moreover, if $G < 0$, the error is asymptotically stable.

Theorem *[14, 39] If there exist gain matrices C_i for $i = 1, 2$, such that the error system (Eq. (13)) can be linearized as $\dot{e}(t) = Ge(t)$ and the matrix $G = A_i - BC_i < 0$, then the error system (Eq. (13)) is asymptotically stable and the slave chaotic system (Eq. (9)) can synchronize the master chaotic system under the fuzzy controller (Eq. (12)).* □

The proof to this theorem can be found in the original source [14].

4 Fuzzy Projective Synchronization

Let be our master system $\dot{x}(t) = f(x(t))$ where $x(t) = (x_1(t), x_2(t), \ldots, x_n(t)) \in \mathbb{R}^n$. And let the slave system be defined as $\dot{y}(t) = g(y(t), u(x(t), y(t)))$ where $y(t) = (y_1(t), y_2(t), \ldots, y_n(t)) \in \mathbb{R}^n$ and $u(x(t), y(t))$ is the controller.

When $\lim_{t \to \infty} \|y(t) - \alpha x(t)\| = 0$ with $\alpha \neq 0$, then the synchronization is referred as projective, and α is called the *scaling factor*.

The master system is given by its fuzzy modelling in Eq. (15).

$$\begin{aligned}
&\text{Rule 1}: \ \mathbf{IF}\ x_1(t)\ \text{is}\ F_1\ \mathbf{THEN}\ \dot{x}(t) = A_1 x(t) \\
&\text{Rule 2}: \ \mathbf{IF}\ x_1(t)\ \text{is}\ F_2\ \mathbf{THEN}\ \dot{x}(t) = A_2 x(t)
\end{aligned} \qquad (15)$$

and the defuzzification process is given in Eq. (16):

$$\dot{x} = \sum_{i=1}^{2} F_i(x_1(t))A_i x(t) \qquad (16)$$

The slave fuzzy system is given by Eq. (17)

$$\begin{aligned}
&\text{Rule 1}: \quad \mathbf{IF}\ x_1(t)\ \text{is}\ F_1 \quad \mathbf{THEN}\ \dot{y}(t) = A_1 y(t) + u(t) \\
&\text{Rule 2}: \quad \mathbf{IF}\ x_1(t)\ \text{is}\ F_2 \quad \mathbf{THEN}\ \dot{y}(t) = A_2 y(t) + u(t)
\end{aligned} \qquad (17)$$

and the final output is presented in Eq. (18).

$$\dot{x} = \sum_{i=1}^{2} F_i(x_1(t))\{A_i y(t) + u(t)\} \tag{18}$$

In projective synchronization, the error is defined as $e(t) = y(t) - \alpha x(t)$. As explained before, by taking the derivative of the error and substituting Eqs. (16) and (18), we can obtain Eq. (19):

$$\dot{e}(t) = \sum_{i=1}^{2} F_i(x_1(t))\{A_i e(t) + u(t)\} \tag{19}$$

By means of parallel distributed compensation (PDC), the fuzzy controller in Eq. (20) is obtained:

Rule 1 :   **IF** $x_1(t)$ is F_1**THEN** $u(t) = -\gamma A_1[y(t) - \alpha x(t)] - \gamma[y(t) - \alpha x(t)]$

Rule 2 :   **IF** $x_1(t)$ is F_2**THEN** $u(t) = -\gamma A_2[y(t) - \alpha x(t)] - \gamma[y(t) - \alpha x(t)]$
$$\tag{20}$$

where γ is a controlling parameter.

The final output of the fuzzy controller is given by Eq. (21):

$$u(t) = \sum_{i=1}^{2} F_i(x_1(t))\{-\gamma A_i[y(t) - \alpha x(t)] - \gamma[y(t) - \alpha x(t)]\} \tag{21}$$

By substituting Eq. (21) in Eq. (19) it is possible to obtain the closed-loop system in Eq. (22).

$$\dot{e}(t) = \sum_{i=1}^{2} F_i(x_1(t))[(1 - \gamma)A_i - \gamma I]e(t) \tag{22}$$

The following theorem and its corollary provide the sufficient conditions to guarantee the projective synchronization of chaotic systems.

Theorem *[17] If there exists a positive definite symmetric constant matrix P and constant $c > 0$, such that $[(1 - \gamma)A_i - \gamma I)]^T P + P[(1 - \gamma)A_i - \gamma I)] \leq -cI$ for $i = 1, 2$, then the equilibrium of the fuzzy system Eq. (22) is globally exponentially stable which implies that Eqs. (15) and (17) can asymptotically achieve projective synchronization.* □

Corollary *[17] Suppose $\{\lambda_j^i\}_{j=1}^{n}$ for $i = 1, 2$ are the eigenvalues of the symmetric matrices $(1 - \gamma)(A_i^T + A_i) - 2\gamma I, i = 1, 2$. If $\max_{1 \leq i \leq 2} \left\{\lambda_j^i\right\}_{j=1}^{n} < 0$, then the equilibrium of the fuzzy system Eq. (22) is globally exponentially stable which implies that Eqs. (15) and (17) can asymptotically achieve projective synchronization.* □

For the proof of these theorem and corollary, the reader is referred to the original source [17].

5   Methodology

Using the fuzzy controller presented in Sects. 3 and 4, we attempt here to apply the complete and projective synchronizations of Chua and Rossler systems to EEG signals where the EEG records act as masters, with either Chua or Rossler systems performing as slave. Figure 3 illustrates the control loop.

5.1   EEG Datasets

In this work we employ data from two publicly available databases. The first EEG dataset is hosted in the *UCI Machine Learning Repository*, it was originally acquired by Henri Begleiter [2] and comes from a large ($n = 122$) study to examine EEG correlates of genetic predisposition to alcoholism. It contains measurements from 64 electrodes located at standard sites (Standard Electrode Position Nomenclature, American Electroencephalographic Association 1990), sampled at 256 Hz (3.9-ms epoch) for 1 second. For the purposes of the current work, a record from the control group with id 337 (subject co2, trial 0 and a single stimulus shown) was selected. In the remaining text, this EGG sample is referred as `data01`.

The second database employed is the Brain Signals and Facial Images dataset presented at Emotion Detection [25]. Emotionally driven physiological signals from both the peripheral (galvanic skin response, respiration and blood volume pressure) and central nervous system (EEG and frontal fNIRS) was acquired. All EEG signals were recorded at 1024 Hz sampling rate except the first session of participant 1 that was recorded at 256 Hz. For the purposes of the current work, record Subject 1, session 1, with id 03082006 was selected. In the remaining text, this EEG record is referred as `data02`.

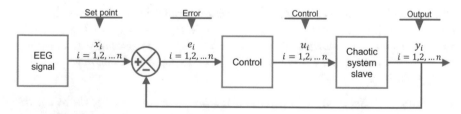

Fig. 3 Schematic depiction of the control used for the synchronization of known chaotic systems to EEG records

5.2 *Processing*

Since both considered slave systems consist of three state variables, three EEG channels were selected. For `data01`, channels x_1 =FP2, x_2 =FC5 and x_3 =CP1 were chosen, which exhibited low correlation (lower than $R^2 = 0.31$). No further filtering was applied to `data01`. For `data02`, channels x_1 =C1, x_2 =C3 and x_3 =Cz were chosen and mean corrected.

In this work the numerical simulations were performed using the following parameters. For the Chua's system: $\alpha = 8.4562$, $\beta = 12.0732$, $\gamma = 0.0052$, $m_0 = -0.1768$, $m_1 = -1.1468$, $d = 1.1$, with initial conditions were $y_0 = [-3.7727, -1.3511, 4.6657]^T$ [8]; for Rossler's system: $a = 0.34$, $b = 0.4$, $c = 4.5$, $d = 10$ and initial condition $y_0 = [1.5, 2.0, 2.0]^T$ [30].

For complete synchronization, the gain matrices must comply with Theorem presented in Sect. 3. The gain matrices were calculated using Matlab LMI toolbox (Mathworks, US). For projective synchronization, the controlling parameters were calculated according to the Corollary shown in Sect. 4.

All synchronization simulations were carried out in Matlab 2017a (Mathworks, US). In all cases, the integration method was Gautschi [6] which has been shown to have better precision than Forward Euler (FE) and requires less evaluations than Runge-Kutta 4 (RK4) [20]. Gautschi's integration method requires declaring the system frequency and integration step. When the system's frequency is unknown, the literature recommends underestimating it [6]. Hence, we underestimated the system's frequencies; we used 1.0 for both Chua and Rosssler systems The integration step was determined using the inverse of the sampling rate. In both cases, the integration step was $1/256$. Despite their popularity, time units may be ambiguous, and therefore we have chosen to report simulations in terms of iterations instead. The total number of iterations is calculated as ($final\ time - initial\ time$)/$integration\ step$. In synchronizations with `data01`, the final time was $1s$ and initial time was $0s$, corresponding to 256 iterations. With `data02`, the final time was $19.53s$ and initial time was $0s$ that corresponds to 5000 iterations.

From the error definition, the complete and the projective synchronizations are equivalent. However, in this work the design of their respective controllers differs. This difference is highlighted in the simulations.

6 Synchronizations with `data01`

Firstly, a set of 4 simulations involving `data01` as the master are considered; these arise from the possible combinations of the two chaotic systems (Chua and Rossler) and the two error functions (complete and projective) with their respective controller designs.

6.1 Chua Complete Synchronization

This simulation comprises data01 and Chua's system as the slave. The gain matrices used for the synchronization are given in Eq. (23) and satisfy the Theorem presented in Sect. 3.

$$C_1 = \begin{pmatrix} 1.34562 & 0 & 0 \\ 0 & -0.5 & 0 \\ 0 & 0 & 0.49479 \end{pmatrix}$$
$$C_2 = \begin{pmatrix} -17.258019 & 0 & 0 \\ 0 & -0.5 & 0 \\ 0 & 0 & 0.49479 \end{pmatrix} \tag{23}$$

Figure 4 presents the phase space of the EEG signals and the synchronized Chua's system response. The plots are accompanied by the traces in each individual state variable as a function of the iteration number. It can be appreciated how the response of the Chua slave follows the main trend of the EEG response while ignoring the quick variations as one would expect from a control capable of dealing with a stochastic phenomenon. Figure 5 shows the evolution of the errors which are trapped within a band around zero (reminiscent of Lyapunov's stability [40]).

The successful synchronization presented above will allow us to now present failed cases and discuss the challenges posed by them.

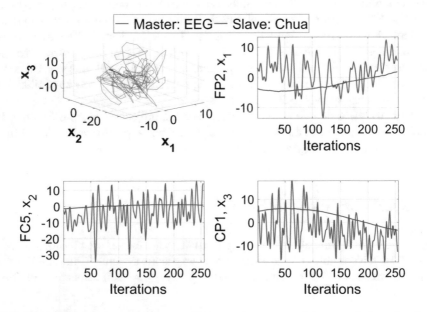

Fig. 4 Phase space and state variable traces. This example corresponds to a complete synchronization using Chua system as a slave over data01

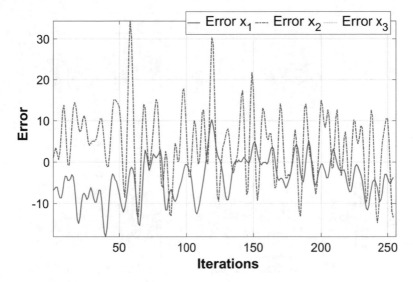

Fig. 5 Synchronization errors for complete synchronization using the Chua system over data01

6.2 Rossler Complete Synchronization

In this complete synchronization the master data comes from data01 while Rossler is used as the slave system. The gain matrices in Eq. (24) satisfy the Theorem shown in Sect. 3.

$$
\begin{aligned}
C_1 &= \begin{pmatrix} 0.5 & 0 & 0 \\ 0 & 0.84 & 0 \\ 0 & 0 & -9.4 \end{pmatrix} \\
C_2 &= \begin{pmatrix} 0.5 & 0 & 0 \\ 0 & 0.84 & 0 \\ 0 & 0 & 10.5 \end{pmatrix}
\end{aligned}
\tag{24}
$$

Figure 6 plots the phase space and state variable traces, and Fig. 7 shows the evolution of the errors. It can be appreciated that synchronization failed in this case, with the first and third state variables failing to converge. This example illustrates how the synchronization of chaotic systems does not follow a simple pick-and-mix recipe and suggests that not any chaotic system can be used a suitable slave candidate.

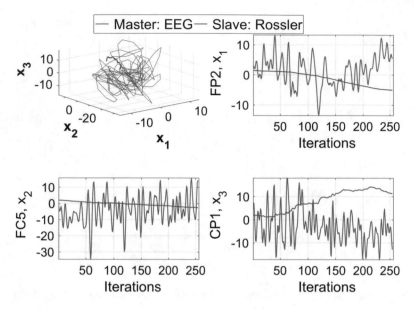

Fig. 6 Phase space and state variable traces. This example corresponds to a complete synchronization using Rossler system as a slave over `data01`

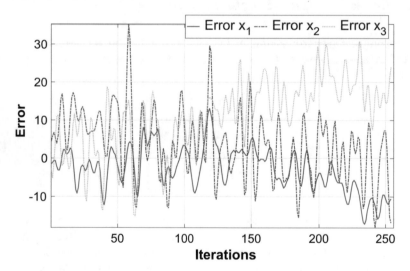

Fig. 7 Synchronization errors for complete synchronization using Rossler system over `data01`

6.3   Chua Projective Synchronization

Projective synchronization generalizes complete synchronization. Here the scaling factor used is $\alpha = 1$, for which complete and projective synchronization should lead to the same error values under analogous situations, but the system dynamic behaviour may differ due to different controller designs.

In this simulation data01 was used as master while Chua served as slave. The controlling parameter was set to $\gamma = 0.9860$.

Results for this simulation are summarised in Figs. 8 and 9. Contrary to our expectations, this synchronization was unsuccessful. Remember that the complete synchronization with Chua was successful over this same EEG record. This example also serves to illustrate how the synchronization of chaotic systems does not follow a simple one-size-fits-all recipe. The same chaotic system may succeed or fail depending on its companion error definition and controller design.

6.4   Rossler Projective Synchronization

The last simulation attempts a projective synchronization using data01 as master and Rossler system as slave. The scaling factor $\alpha = 1$ is maintained while the controlling parameter is set to $\gamma = 0.93$.

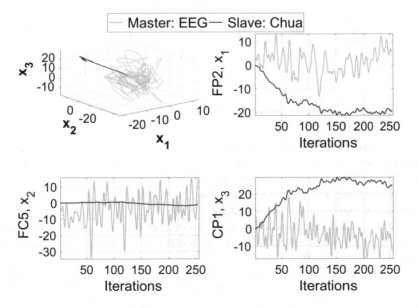

Fig. 8 Phase space and state variable traces. This example corresponds to a projective synchronization using Chua system as a slave over data01

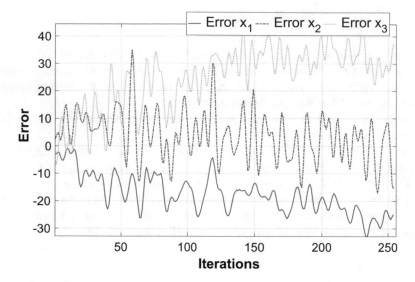

Fig. 9 Synchronization errors for projective synchronization using Chua system over data01

Analogously to previous cases, we show the phase space and state variables in Fig. 10 and the error traces in Fig. 11. While synchronization might not have been achieved within the simulated timeframe, it also seems that the behaviour of the Rossler slave with the projective synchronization improves on the results obtained with the complete counterpart. Only the third state variable exhibits a circumstantial departure from the master but by the end of the simulation, it looks like the error may still be confined.

7 Synchronizations with data02

The second set of 4 simulations are analogous to the previously described but replacing data02 as master instead of data01. While in data01 the range of all state variables is approximately similar in magnitude, data02 presents the additional difficulty of the range of the state variables being prominently different. Since the partial error in each state variable contributes equally to total error, it is expected to be more difficult to achieve synchronization using the complete and projective synchronizations. It is beyond the scope of this study to explore other error functions such as the generalized projective.

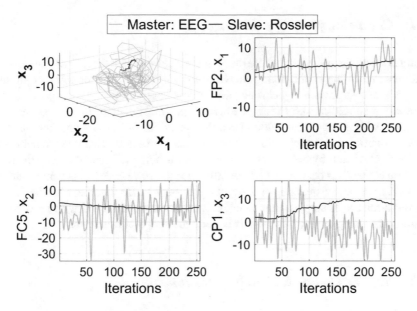

Fig. 10 Phase space and state variable traces. This example corresponds to a projective synchronization using Rossler system as a slave over `data01`

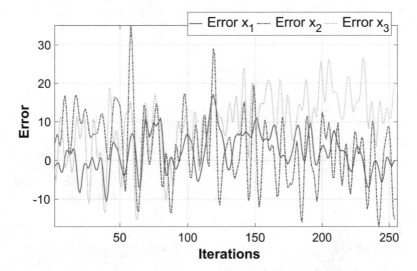

Fig. 11 Synchronization errors for projective synchronization using Rossler system over `data01`

7.1 Chua Complete Synchronization

Chua is used as slave system while the gain matrices used are those found in Eq. (23), the outcome of the simulation is shown in Figs. 12 and 13. Although it is possible to speak of a successful synchronization in this case, the insight that provides this simulation should not be oversighted. As expected, the error on the second state variable, with a larger range dominates the behaviour of the synchronization. Note how the partial errors over state variables x_1 and x_3 roughly resemble the behaviour of the residuals left by the partial error on state variable x_2. Recall that only the mean of these signals has been removed but otherwise the raw EEG data is being used. No effort at filtering, normalizing or detrending has been made. The clear trend exhibited by the error traces suggest that synchronization can be achieved with an adequate signal processing strategy, but this is beyond the scope of this study.

7.2 Rossler Complete Synchronization

Dataset `data02` is used as master while Rossler is used as slave system and the gain matrices used are those found in Eq. (24), the outcome of the simulation is shown in Figs. 14 and 15. The divergent pattern of the error observed in this simulation is difficult to interpret. We observe a circumstantial reduction of the error around

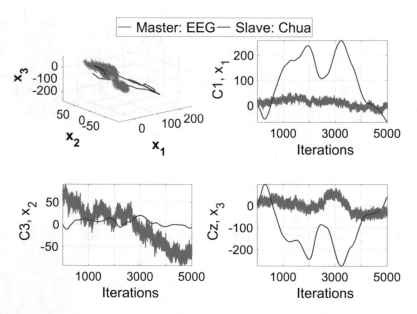

Fig. 12 Phase space and state variable traces. This example corresponds to a complete synchronization using Chua system as a slave over `data02`

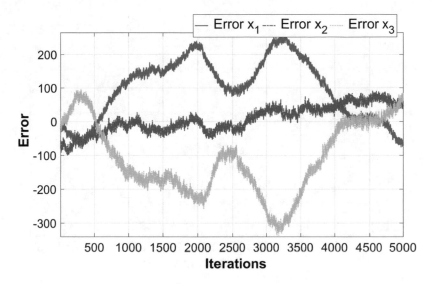

Fig. 13 Synchronization errors for complete synchronization using Chua system over `data02`

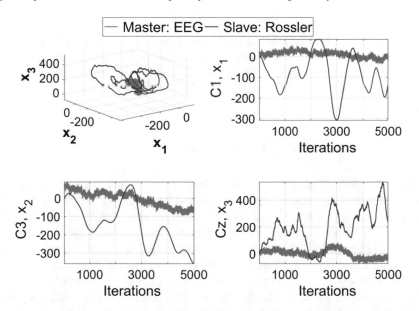

Fig. 14 Phase space and state variable traces. This example corresponds to a complete synchronization using Rossler system as a slave over `data02`

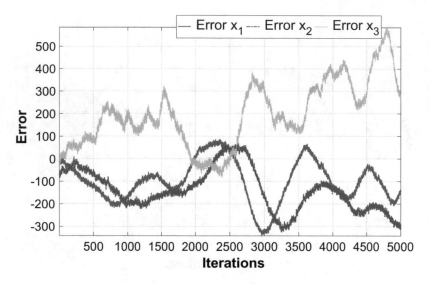

Fig. 15 Synchronization errors for complete synchronization using Rossler system over data02

iteration 2300–2400. Since this is not accompanied by any striking feature on the master trace, it is hypothesized that perhaps this corresponds to a natural oscillation in the slave system. Should this be the case, longer simulations would be expected to exhibit irregular cycles of error increases and decreases.

7.3 Chua Projective Synchronization

Dataset data02 is used as master while Chua is used as slave system. Scaling factor α and the controlling parameter γ are set to 1 and 0.9860 respectively, the outcome of the simulation is shown in Figs. 16 and 17. Another failed attempt is observed when using Chua's system. Although the slave system state variable x_2 mostly follows the reference of channel C3-the one with the larger range, no adequate synchronization seems to occur. The final overcompensation imposed on the state variables x_1 and x_3 seem to be driven by the change in trend in C3 around iteration 2600–2700, and almost immediately followed by the "bump" in Cz from 2700 until 3300. Only in the final iterations, as the trend in C3 hints of another change in trend, the error appears to recover slightly.

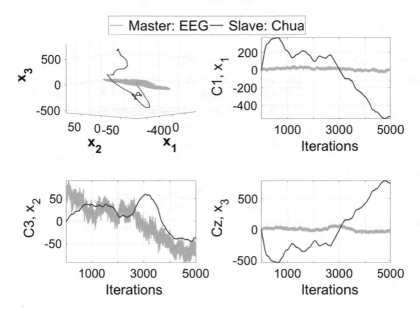

Fig. 16 Phase space and state variable traces. This example corresponds to a projective synchronization using Chua system as a slave over `data02`

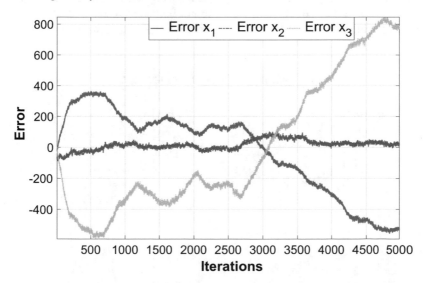

Fig. 17 Synchronization errors for projective synchronization using Chua system over `data02`

7.4   Rossler Projective Synchronization

Finally, data02 is used as master while Rossler is used as slave system. Scaling
factor α and controlling parameter γ are set to 1 and 0.93 respectively, the outcome
of the simulation is shown in Figs. 18 and 19. The results of this simulation seem
to support the previous observation of the variable with the largest range acting as
driving force. The error is clearly dominated by C3 and Rossler slave response mostly
follows this reference. The mid-simulation behaviour, similar to the case of complete
synchronization, may be explained by a natural oscillation in the slave system and
while this possibility is not explored in this work, if this were the case, then it is
suggested that an adequate strategy for the synchronization could be pairing the
slave system's natural oscillations with the EEG energy band dominant oscillations
so that they respond coherently.

8   Discussion

EEG records exhibit chaotic behaviour and thus could be susceptible to being mod-
elled using chaotic systems, for example, through synchronization. Although it is
possible for well-studied chaotic models to not be adequate for this task, it is still
worth exploring their suitability. This comparative study attempts to highlight those

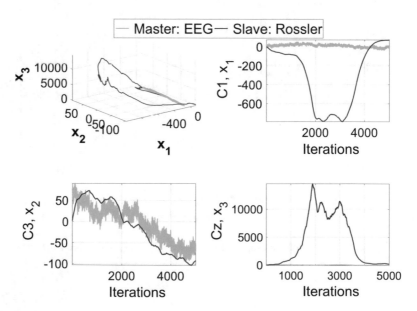

Fig. 18 Phase space and state variable traces. This example corresponds to a projective synchronization using Rossler system as a slave over data02

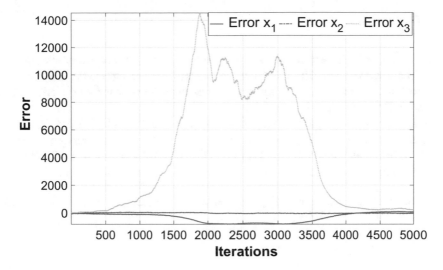

Fig. 19 Synchronization errors for projective synchronization using Rossler system over `data02`

known chaotic systems, under specific conditions that have to be analysed in each case, are potentially useful. But at the same time, it also emphasizes how challenging this task is; synchronizing EEG and known chaotic systems do not follow a straightforward recipe. Fuzzifying the chaotic systems facilitate the task -by only requiring recalculating the gain matrices or the controlling parameter, but it remains non-trivial.

When synchronization is successful, the chaotic system appears to be modelling the central tendency of the EEG record, disregarding the sample variations. Errors are therefore in the same range as the EEG oscillations. It is possible that by oversampling the EEG and reducing the integration step, to force the chaotic slave to partially accommodate the sample variations, but this remains to be tested.

The controllers used requires that the theorems presented in Sects. 3 and 4, respectively, be fulfilled to have successful synchronizations, assuming that the mathematical models of the master and slave systems are available. However, in this work we do not know the mathematical model of the master system, so checking the above constraints is not immediate. We have observed that, in our case, not all the synchronizations carried out were successful, which is something that needs further investigation.

In this study, it was decided to use three channels chosen based on their low correlations. However, if we want to preserve more information of the EEG, an alternative is to use the first $k = 3$ temporal components as established with principal component analysis.

9 Conclusions

We have shown both the feasibility and at the same time the difficulty in synchronizing known chaotic systems with EEG records as a way to characterize these latter. The contrasting hits and misses in our simulations underscore that there is still much more work ahead in this field.

As future work, we propose to perform dynamic analysis of the EEG signals to determine their chaotic properties such as Lyapunov exponents, Kolmogorov-Sinai entropy, bifurcation analysis, etc. From this analysis, we intend to look for chaotic systems with similar properties to describe the EEG signals more adequately.

Acknowledgements The author J. Zaqueros-Martinez was supported by the Mexican Research Council (CONACYT no. 776446).

References

1. Ahmad, S., Mohd Ali, B., Wan Adnan, W.A.: Technical issues and challenges of biometric applications as access control tools of information security. Int. J. Innov. Comput. Inf. Control **8**, 7983–7999 (2012)
2. Begleiter, H.: Statistical mechanics of neocortical interactions: canonical momenta indicators of electroencephalography. Phys. Rev. E 4578–4593 (1997)
3. Chua, L.O.: The genesis of chua's circuit. Archiv fur Elektronik und Uebertragungstechnik **46**(4), 250–257 (1992)
4. Chua, L.O.: A zoo of strange attractors from the canonical chua's circuits. Technical Report UCB/ERL M92/87, EECS Department, University of California, Berkeley (1992)
5. Dayan, P., Abbott, L.F.: Theoretical Neuroscience: Computational and Mathematical Modeling of Neural Systems. The MIT Press (2001)
6. Gautschi, W.: Numerical integration of ordinary differential equations based on trigonometric polynomials. Numer. \ Math. **3**:381–397 (1961)
7. Gui, Q., Jin, Z., Xu, W.: Exploring eeg-based biometrics for user identification and authentication. In: 2014 IEEE Signal Processing in Medicine and Biology Symposium (SPMB), pp 1–6 (2014)
8. Kiseleva, M.A., Kudryashova, E.V., Kuznetsov, N.V., Kuznetsova, O.A., Leonov, G.A., Yuldashev, M.V., Yuldashev, R.V.: Hidden and self-excited attractors in Chua circuit: synchronization and SPICE simulation. Int. J. Parallel, Emerg. Distrib. Syst. **33**(5), 513–523 (2018)
9. Korn, H., Faure, P.: Is there chaos in the brain? II. experimental evidence and related models. Comptes Rendus Biol. **326**(9):787–840 (2003)
10. Kumar, S., Singh, C., Prasad, S.N., Shekhar, C., Aggarwal, R.: Synchronization of fractional order Rabinovich-Fabrikant systems using sliding mode control techniques. Arch. Control Sci. **29**(2), 307–322 (2019)
11. Kuznetsov, N., Leonov, G., Vagaitsev, V.: Analytical-numerical method for attractor localization of generalized chua's system*. In: 4th IFAC Workshop on Periodic Control Systems IFAC Proceedings Volumes, vol. 43, no. 11, pp. 29–33 (2010)
12. Li, Z.: Fuzzy Chaotic Systems: Modeling, Control, and Applications. Springer, Berlin, Heidelberg (2006)
13. Lian, K.Y., Chiang, T.S., Chiu, C.S., Liu, P.: Synthesis of fuzzy model-based designs to synchronization and secure communications for chaotic systems. IEEE Trans. Syst. Man Cybern. Part B: Cybern. **31**(1), 66–83 (2001)

14. Lian, K.Y., Chiu, C.S., Chiang, T.S., Liu, P.: LMI-based fuzzy chaotic synchronization and communications. IEEE Trans. Fuzzy Syst. **9**(4), 539–553 (2001)
15. Lin, T.-C., Lee, T.-Y.: Chaos synchronization of uncertain fractional-order chaotic systems with time delay based on adaptive fuzzy sliding mode control. IEEE Trans. Fuzzy Syst. **19**(4), 623–635 (2011)
16. Mamdani, E.: Application of fuzzy algorithms for control of simple dynamic plant. In: Proceedings of the Institution of Electrical Engineers, vol. 121, pp. 1585–1588 (1974)
17. Nian, Y., Zheng, Y.: Generalized projective synchronization of chaotic systems based on takagi-sugeno fuzzy model. In: 2010 Seventh International Conference on Fuzzy Systems and Knowledge Discovery, vol. 3, pp. 1291–1295 (2010)
18. Nunez, P.L., Srinivasan, R.: Electric Fields of the Brain: the Neurophysics of EEG. Oxford University Press, USA (2006)
19. Ouannas, A., Bendoukha, S., Volos, C., Boumaza, N., Karouma, A.: Synchronization of fractional hyperchaotic rabinovich systems via linear and nonlinear control with an application to secure communications. Int. J. Control Autom. Syst. **17**(9), 2211–2219 (2019)
20. Pano-Azucena, A.D., Tlelo-Cuautle, E., Rodriguez-Gomez, G., de la Fraga, L.G.: FPGA-based implementation of chaotic oscillators by applying the numerical method based on trigonometric polynomials. AIP Adv. **8**(7), 75217 (2018)
21. Pecora, L.M., Carroll, T.L., Johnson, G.A., Mar, D.J., Heagy, J.F.: Fundamentals of synchronization in chaotic systems, concepts, and applications. Chaos: Interdiscip. J. Nonlinear Sci. **7**(4):520–543 (1997)
22. Pham, V.-T., Volos, C., Vaidyanathan, S., Wang, X.: A chaotic system with an infinite number of equilibrium points: dynamics, horseshoe, and synchronization. Adv. Math., Phys (2016)
23. Pijn, J.P., Van Neerven, J., Noest, A., Lopes da Silva, F.H.: Chaos or noise in eeg signals; dependence on state and brain site. Electroencephalogr. Clin. Neurophysiol. **79**(5), 371–381 (1991)
24. Rössler, O.: An equation for continuous chaos. Phys. Lett. A **57**(5), 397–398 (1976)
25. Savran, A., Ciftci, K., Chanel, G., Mota, J.C., Viet, L.H., Sankur, B., Akarun, L., Caplier, A., Rombaut, M.: Emotiondetection in the loop from brain signals and facial images. In: eNTERFACE (2006)
26. Shahzad, M., Pham, V.T., Ahmad, M.A., Jafari, S., Hadaeghi, F.: Synchronization and circuit design of a chaotic system with coexisting hidden attractors. Eur. Phys. J. Spec. Top. **224**(8), 1637–1652 (2015)
27. Sterratt, D., Graham, B., Gillies, A., Willshaw, D.: Principles of Computational Modelling in Neuroscience. Cambridge University Press (2011)
28. Sambas, A., He, S., Liu, H., Vaidyanathan, S., Hidayat, Y., Saputra, J.: Dynamical analysis and adaptive fuzzy control for the fractional-order financial risk chaotic system. Adv. Differ. Equ. **2020**(1), 674 (2020)
29. Takagi, T., Sugeno, M.: Fuzzy identification of systems and its applications to modeling and control. IEEE Trans. Syst. Man Cybern. SMC **15**(1), 116–132 (1985)
30. Tanaka, K., Ikeda, T., Wang, H.O.: A unified approach to controlling chaos via an lmi-based fuzzy control system design. IEEE Trans. Circ. Syst. I: Fundam. Theory Appl. **45**(10), 1021–1040 (1998)
31. Thomas, P.A., Mathew, M.K.: A broad review on non-intrusive active user authentication in biometrics. J. Ambient Intell. Human, Comput (2021)
32. Trappenberg, T.P.: Fundamentals of Computational Neuroscience. Oxford University Press (2010)
33. Wang, B., Cao, H., Wang, Y., Zhu, D.: Linear matrix inequality based fuzzy synchronization for fractional order chaos. Math. Probl., Eng (2015)
34. Wang, J.S.: Exploring biometric identification in fintech applications based on the modified tam. Financ. Innov. **7**(1), 42 (2021)
35. Wang, Y.-W., Guan, Z.-H., Wang, H.O.: Lmi-based fuzzy stability and synchronization of chen's system. Phys. Lett. A **320**(2), 154–159 (2003)

36. Wayman, J.L., Jain, A.K., Malton, D., Maio, D.: Biometric Systems: technology. Design and Performance Evaluation. Springer, London (2005)
37. Weng, T., Yang, H., Gu, C., Zhang, J., Small, M.: Synchronization of chaotic systems and their machine-learning models. Phys. Rev. E **99**(4), 1–7 (2019)
38. Zhang, G., Wu, F., Wang, C., Ma, J.: Synchronization behaviors of coupled systems composed of hidden attractors. Int. J. Modern Phys. B **31**(26), 1750180 (2017)
39. Zhang, H., Liao, X., Yu, J.: Fuzzy modeling and synchronization of hyperchaotic systems. Chaos Solitons Fractals **26**(3), 835–843 (2005)
40. Zhang, H., Liu, D., Wang, Z.: Controlling Chaos: suppression. Synchronization and Chaotification. Springer, London (2009)
41. Zheng, G., Liu, L., Liu, C.: Hidden coexisting attractors in a fractional-order system without equilibrium: analysis, circuit implementation, and finite-time synchronization. Math. Probl., Eng (2019)
42. Zheng, Y., Shang, L.: Generalized projective synchronization of takagi–sugeno fuzzy drive-response dynamical networks with time delay. In: Sun, Z., Deng, X., (eds.), Proceedings of 2013 Chinese Intelligent Automation Conference. Springer, Berlin, Heidelberg, pp 119–126 (2013)
43. Zhou, P., Zhu, P.: A practical synchronization approach for fractional-order chaotic systems. Nonlinear Dyn. **89**(3), 1719–1726 (2017)

Secure Communication Scheme Based on Projective Synchronization of Hyperchaotic Systems

Freddy Alejandro Chaurra-Gutierrrez, Gustavo Rodriguez-Gomez,
Claudia Feregrino-Uribe, Esteban Tlelo-Cuautle,
and Omar Guillen-Fernandez

Abstract This chapter develops a secure communication scheme based on the synchronization of two hyperchaotic systems, including key statistical, dynamical, and security analyses. First, the scheme is based on the projective synchronization between two hyperchaotic systems applied to image encryption. The encryption process comprises the generation and discretization of chaotic sequences, the transformation of the sequence elements into the image domain, and the modulo operation as encryption transformation. Second, the reliability and robustness of the proposed scheme are evaluated through the primary dynamical, statistical, and security analyses such as spectral and dynamical complexity, sensitivity, unpredictability, chaotic range, elements distribution, and randomness characteristics, key space definition, and resistance to distortion analysis, return map analysis, differential attack, decryption, and brute force attacks. Finally, the experimental results give evidence of the feasibility and robustness of the chaotic scheme. They show that using high-dimensional chaotic systems such as hyperchaotic systems combined with projective synchronization improves the confusion and diffusion properties, increases the key space, and limits the effectiveness of decryption, distortion, differential, generalized and adaptive synchronization, return map and brute force attacks.

F. A. Chaurra-Gutierrrez · G. Rodriguez-Gomez (✉) · C. Feregrino-Uribe · E. Tlelo-Cuautle ·
O. Guillen-Fernandez
INAOE, Puebla, Mexico
e-mail: grodrig@inaoep.mx

F. A. Chaurra-Gutierrrez
e-mail: chaura@inaoep.mx

C. Feregrino-Uribe
e-mail: cferegrino@inaoep.mx

E. Tlelo-Cuautle
e-mail: etlelo@inaoep.mx

© The Author(s), under exclusive license to Springer Nature Switzerland AG 2022   109
A. A. Abd El-Latif and C. Volos (eds.), *Cybersecurity*, Studies in Big Data 102,
https://doi.org/10.1007/978-3-030-92166-8_6

1 Introduction

The rapid development of modern communication systems, coupled with improved computing performance, has enabled the transmission of more and more digital information in social networks, academic applications, telemedicine, and satellite communications, mainly in image format [1, 2]. This multimedia content can carry sensitive, or private information [1, 3]. Thus, research on protecting such information has become a highly relevant topic [1, 3, 4]. However, due to the large amount of data and the high correlation in images, protect these multimedia contents is a challenge between efficiency and security.

Among the new proposals to address information security is chaos-based encryption schemes, which take advantage of the intrinsic properties of chaotic systems, such as extreme sensitivity to small perturbations, high unpredictability, ergodicity, and pseudo-randomness [4–6].

In secure communications, a great variety of secure chaos-based synchronization schemes has been implemented based on three elementary approaches: parameter modulation [7], where the signal or message modulate the chaotic system parameters, chaotic masking [8–10], in which the message is added directly to the chaotic carrier, and chaotic shift keying [11], used to carry binary signals through two or more different chaotic attractors. Besides, some works show that these basic methods are insecure to various attacks [4, 12–14], where the main issue is the existence of patterns in the transmitted signal that could reveal the hidden message features or the chaotic system dynamics [5, 14, 15]. Both consequences are undesirable and must be limited to ensure the robustness of the schemes.

Chaos-based secure communication proposals generally are insecure, so develop and protect communication schemes remains a real challenge [12, 14–18]. In general, it is difficult to evaluate the security levels of the new proposals in this field, leaving an open door for different attacks. The most common security issues are weak key definition, wrong selection of the chaotic system, lack of implementation details, and shallow analysis of the dynamical, computational, and security aspects [12, 14–18]. Thus, many proposals do not provide deep analysis and fail to explain the principal characteristics fundamental to all secure communication schemes [5, 12, 14–22]. Therefore, it is necessary a deep analysis to improve the levels of security and reduce attacks.

There are several studies in chaotic secure communication, many of which have demonstrated significant advances in dealing with security drawbacks. Some proposed solutions use hyperchaotic or fractional systems [14–16, 23], using different synchronization types such as generalized, projective, complex, or impulsive synchronization [14–16, 24, 25], and modifications to the master/slave configuration such as cascade, dual, and combined synchronization have been presented [14, 26–28]. Furthermore, some research analyzes the most common errors and wrong design/analysis practices in the field. These studies present different solutions and propose basic development guidelines to overcome security deficiencies [14, 15, 17]. Besides, the research in chaotic cryptanalysis taking advantage of the lack of analysis and security weaknesses [21, 29–37].

In recent years, many image encryption schemes based on chaotic systems have moved from low-dimensional systems to high-dimensional, or more complex systems, such as hyperchaotic ones [2, 3, 15, 16, 23, 38]. It is due to the high sensitivity to slight changes, the high level of unpredictability, the more complex orbits, and the wide bandwidth of hyperchaotic systems relative to low-dimensional chaotic systems. These properties make hyperchaotic systems suitable for cryptographic applications [1–3, 14]. Li et al. [3] proposed a secure image encryption based on hyperchaotic Lorenz system and hash function. They used a permutation-diffusion process to encrypt the message. Xu et al. [2] developed a secure scheme using a new hyperchaotic map, compressive sensing, and SHA-512 hash to generate the secret key. Zhang et al. [1] designed an image encryption algorithm based on a hyperchaotic system and variable-step Josephus problem, where the chaotic sequences are inputs into the Josephus function to scramble the image elements. Zhao and Ren [6] presented a secure image algorithm using a hyperchaotic Chen system generated by linear time-delay feedback method. Ahmad et al. [4] performed a security analysis and developed a color image cryptosystem based on permutation only cipher and synchronized hyperchaotic system. Li and Zhang [39] proposed an hyperchaotic image encryption via multiple bit permutation and diffusion determined by chaotic sequences. Naim et al. [40] presented a satellite image encryption algorithm using the Linear Feedback Shift Register generator, SHA-512, hyperchaotic system, and Josephus problem. Yang et al. [41] developed a new hyperchaotic Liu system applied to image encryption. They used chaotic sequences and DNA sequence operations to designed the cryptosystem. Bouridah et al. [42] addressed the hyperchaotic synchronization problem via static error feedback applied to image encryption. The previous works present appropriate statistical characteristics. However, they do not perform the robustness analysis against some of the known cryptographic attacks. Therefore, they cannot completely guarantee the security of the encryption scheme.

This chapter addresses the problem of modified projective synchronization between two identical hyperchaotic Qi systems in a master-slave configuration and its application to image encryption. First, based on Lyapunov stability theory, we design an active control to synchronize the trajectories of the hyperchaotic systems by a scaling factor. Second, we compute the secret key of the proposed scheme using scaling factors, initial conditions, and system parameters, where the unpredictability of the scaling factors improves the key space. Third, we generate a chaotic sequence with pseudorandom properties using the master system. Finally, we encrypt the original message using the chaotic sequence with a One-Time-Pad structure, which improves the robustness of the cryptosystem due to the higher degree of unpredictability of the system and the security features of the One-time-Pad.

We analyze dynamical properties such as Lyapunov exponents, stability points, parameters number, chaotic range, Kolmogorov-Sinai entropy, and separation time of the hyperchaotic system. The purpose of this analysis is to characterize the system properties in security applications. In addition, we perform an extensive analysis not only considering statistical tests as a security measure but also ensuring robustness against different known cryptographic attacks.

Statistical tests include analysis of elements distribution and uniformity, corre-lation test, and sensitivity characteristics related to any cryptosystem's diffusion, confusion, and randomness properties. We show that the chaos-based scheme has suitable statistical features for encryption applications. The security tests consider key space, sensitivity, distortion, differential, return map, parameter estimation, and known/chosen plaintext analysis on the scheme. These tests increase the robustness against the following attacks: brute force, parameter estimation, decryption, adap-tive and generalized synchronization, distortion, differential attack, return map, and known/chosen plaintext. All the performed analyses should be included as a gen-eral checklist for designing secure chaos-based communication schemes. Finally, we compare the proposed encryption scheme with other works using hyperchaotic systems for image encryption.

This chapter is organized as follows: Sect. 2 introduces basic theory of cryptog-raphy and chaotic systems. Section 3 presents the proposed synchronization and encryption methodology. Section 4 analyzes the security properties of the scheme. The last section concludes the chapter and presents the future work.

2 Background

2.1 Cryptography

The primary purpose of cryptography is to protect information from unauthorized access by third parties through encryption processes that make data incomprehensible to ensure some security-related objectives. Some of these objectives are confiden-tiality, data integrity, message, and identity authentication, non-repudiation, and key exchange [43].

Encryption schemes are the main cryptographic tools used to provide information security. These can be characterized through the following five sets [43, 44]:

- Unencrypted messages (*plaintext*), m.
- Encrypted messages (*cyphertext*), m_c.
- Secret key space, k_i.
- Encryption transformation, e.
- Decryption transformations, d.

The encryption process assigns each element of the key space an encryption function with a unique inverse transformation. Figure 1 shows the basic encryption scheme.

Note: The security of the encryption schemes is based exclusively on the lack of knowledge of the secret key by a third party or attacker so that if the key is not known, no bit of information can be decrypted or read.

Encryption is divided into symmetric and asymmetric. Symmetric uses the same key to encrypt and decrypt, and the encryption process is performed on a block-by-

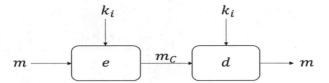

Fig. 1 General encryption/decryption scheme

block or bit-by-bit basis [44]. On the other hand, asymmetric uses different secret keys for the encryption and decryption process. Some of the most popular encryption algorithms are RSA (asymmetric), RC4, A5/1 (block symmetric), AES, and One-Time Pad (bit symmetric) [43, 44].

2.1.1 One-Time Pad

The One-time pad is the most secure cipher. It guarantees *perfect security*, and it is a bit-by-bit cipher (stream cipher), where we encrypt the message with a stream of truly random keys through an 'exclusive OR' (XOR) operation [44, 45].

If the following preconditions are satisfied, it is possible to prove that a stream cipher is unbreakable [44, 45]:

- The key must be as long as the message.
- The key must be truly random.
- The key must only be used once.

The main issues with this encryption system are the difficulty of generating truly random keys, guaranteeing their unique use, and the transmission and storage of the secret key.

Encryption/decryption process: One-time pad takes a random key, k, and a message, m, and produces an encrypted message, m_c, through XOR operation, defined as [44, 45].

$$m_c = m \otimes k \tag{1}$$

where m_c, m, and k are bit strings of the same length. The encryption process is similar to encryption, $m = m_c \otimes k$. Figure 2 shows the encryption scheme.

The following steps describe the encryption/decryption and transmission process [45].

- Generate a random secret key, k, with the same length as the message, m, and transmit it to the receiver.
- The random key is combined bit to bit with the message using the XOR operation.
- The encrypted message is transmitted to the receiving side and decrypted using the same key.
- Both sender and receiver destroy keys are after use.

Fig. 2 One-time pad scheme

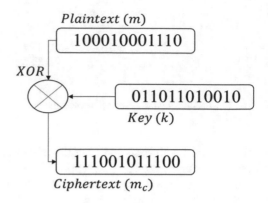

Plaintext (m)

100010001110

XOR

011011010010

Key (k)

111001011100

Ciphertext (m_c)

2.1.2   Security: Design Criteria

Different aspects must be considered when designing secure schemes to guarantee security and compliance with the cryptographic goals (privacy, data integrity, authentication, or non-repudiation) [43, 44]. These aspects are statistical properties, key space definition, computational and implementation aspects, security-related aspects, and cryptanalysis [14, 15, 43].

Statistical properties: An appropriate secure communication system must allow a transformation from the space of messages to the space of encrypted messages so that the corresponding encrypted message is independent of the information characteristics of the message or, if such a dependency exists, it must be of high complexity. This characteristic is expressed through the cryptographic properties of confusion, and diffusion [14, 18, 44].

The confusion property increases the complex relationship between the key and the encrypted message, thus eliminating redundancies and statistical patterns. The diffusion property aims to distribute the information message and the secret key information over the elements of the encrypted message [14, 18]. The features that increase the levels of diffusion and confusion are listed below [14, 15, 17, 18, 20, 43, 44].

- Uniform distribution of values or bits in the encrypted message.
- Entropy close to the maximum coding value.
- Minimal or no correlation between the message and the encrypted message (statistical independence).
- Minimal or no correlation between the elements of the encrypted message.
- High sensitivity to slight variations in the message and the secret key.
- Encrypted message as a set of truly random values.

Key space: The most important feature of any cryptographic system is the definition of the key space of the system, Kerckhoff's principle. It includes the characterization of the key space, the definition of the key components, and the relationship between its elements [14, 15]. The main characteristics that the key of a cryptographic system must have are as follows [14–16, 43, 44, 46].

- Complete definition of the key generation and key selection process.
- A precise definition of the secret key space. If this space is not defined, the same encrypted message could be generated with two different keys.
- The key space of the encryption scheme must be large enough to prevent a brute force attack. Therefore, the size is expected to be larger than 2^{256}.
- The system key must be highly sensitive to small changes in its elements. A slight variation should produce an encrypted message with at least a 90% change in its elements.
- The elements of the key must not be correlated.
- The key must not be recoverable through information in the message or encrypted message.

Computational and implementation aspects: Each of these aspects must be specified in detail so that a third party can replicate and analyze the cryptographic system. The computational and implementation aspects are listed below [14, 15, 18, 43, 44, 47].

- All operations involved in the encryption and decryption process must be detailed.
- Algorithm encryption/decryption time.
- The numerical integration method should be selected so that there is no degradation in the efficiency of the scheme and that it guarantees the preservation of the chaotic dynamical properties of the system for long periods.
- Encryption algorithm implementation problems of software or hardware.
- Propagation error analysis.

Security and cryptanalysis issues: The measure of the quality of a cryptographic system is based on the ability to withstand attempts by a third party (attacker) to obtain information about a hidden message or secret key. This measure is called security and is evaluated by different attacks that attempt to break the system [18, 44]. Therefore, all secure communication systems should always be evaluated by some basic and specific security analysis covering some of the best-known and typical cryptographic tests. These analyses help to accumulate evidence to support the security levels of the complete scheme [14–16, 18, 43, 44, 48–50].

- It should evaluate the resistance of the algorithm to different attack scenarios such as Ciphertext-only, Known-plaintext, Chosen-plaintext, and Chosen-ciphertext attacks.
- The best-known attack is the exhaustive key search or brute force attack. This attack takes advantage of the limitations on the size of the system key space; that is, the effectiveness of the brute force attack is proportional to the key space size.
- A noise distortion analysis should be performed. Generally, the schemes are designed for an ideal transmission channel without distortion, but in practice, some variations and perturbations are referred to as the transmission channel. Furthermore, an interesting attack is related to distorting the encrypted message so

that the receptor cannot retrieve the original message. Therefore, the resistance to slight variations or distortions in the encrypted message should be evaluated.

- Many specific chaotic attacks have been proposed in the literature, which explores and exploits the dynamic deficiencies of the selected chaotic system. The main analyzes and attacks are frequency spectrum analysis and filtering techniques, short-time period analysis, correlation and autocorrelation, parameter estimation, generalized and adaptive synchronization analysis, return map analysis and brute force attacks. Therefore, a careful analysis should be performed to chaotic specific attacks.

2.2 Chaotic Systems

Chaotic dynamics begins with Henri Poincaré and his study of the stability of celestial systems, such as the solar system. Poincaré showed that three bodies under gravitational forces describe very complicated orbits. This type of motion has a more complex aperiodic behavior than conventional motion. Moreover, these systems are highly dependent on initial conditions, making long-term predictions impossible [51–53].

For a long time, chaotic systems were considered problematic and of limited usefulness due to their erratic behavior, difficult to predict and control [52, 53]. However, more powerful computers allowed the development of different experiments to analyze and obtain a deeper insight into nonlinear systems. For example, such experiments allowed Edward Lorenz to discover chaotic motion in atmospheric convection [52, 54]. He found that the solutions are always aperiodic and that slight differences in the initial conditions generate different behavior. The Lorenz equations of atmospheric convection is given by:

$$\dot{x}_1 = \sigma(x_2 - x_1)$$
$$\dot{x}_2 = x_1 (\rho - x_3) - x_2 \qquad (2)$$
$$\dot{x}_3 = x_1 x_2 - \beta x_3$$

where σ, ρ, and β are the system parameters that determine the chaotic regions of the system [54].

Figures 3 and 4 show the time series and phase-space representation of the system, where the trajectories evolve as a random mixture. Also, Figure 5 shows the sensitivity to a small perturbation for the state variable x of the system under the initial conditions $x(0) = 0.1$ and $x(0) = 1.0 \times 10^{-6}$.

Also, a few years later, researchers found that chaos is naturally present in chemistry, mathematics, biology, physics, economics, weather prediction, and engineering. Therefore, efforts have been made to develop mathematical and engineering tools to control and use chaos positively due to its pseudo-random properties [52–54].

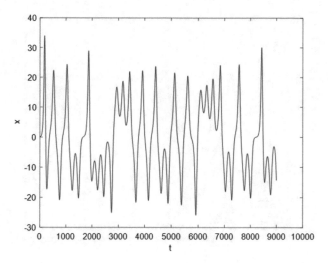

Fig. 3 Time evolution of the state variable x

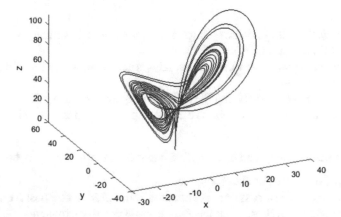

Fig. 4 Phase space: Lorenz system

There is no universally accepted definition of chaos, but it can be defined based on its main characteristics, such as non-periodic, non-linear, deterministic, and extremely sensitive dependence to initial conditions [52, 54]. Thus, the chaotic structure of the attractor is characterized by means:

- *Lyapunov exponents* measures the exponential divergence of the system trajectories.
- *Entropy* measures the unpredictability of the chaotic system.
- *Kaplan-York dimension* characterizes the dimension and structure of the attractors of the system.

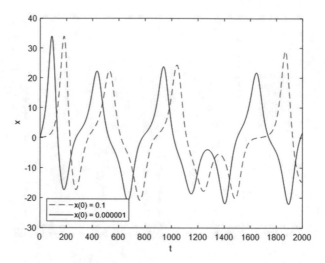

Fig. 5 Time evolution to different initial conditions, $x(0) = 0.1$ and $x(0) = 1.0 \times 10^{-6}$

- *Autocorrelation* allows determining whether a variable is similar to itself at any instant of time (periodicity).
- *Bifurcation diagrams* are qualitative analyses of system behavior in response to parameter changes.
- *Periodigrama* allows analyzing the spectral frequency of the system.
- *Phase space* enables us to observe the complex and irregular behavior of the system.

Also, chaotic systems can be classified based on the following properties [52]:

- *Domain*. Continuous or discrete space.
- *Lyapunov Exponents*. A system is chaotic if it has at least one positive Lyapunov exponent and hyperchaotic if it has two or more positive exponents.
- *Derivative order*. Integer derivative or fractional derivative.
- *Attractors*. Hidden or self-excited attractors.

2.3   Chaotic Synchronization

In our case, the synchronization process refers to the phenomenon where two chaotic systems with slightly different initial conditions are synchronized, such that they converge towards the same trajectory after a specific time interval. However, this phenomenon was initially thought unrelated to chaotic systems due to its sensitivity to system parameters and its behavior similar to random noise [52]. Pecora and Carroll showed that it is possible to synchronize two identical chaotic systems (Master-Slave) if and only if the conditional Lyapunov exponents of the response subsystem are all

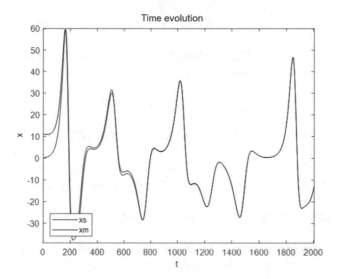

Fig. 6 Time evolution of synchronization for the variable x of the master/slave system

negative for a particular input [52, 55]. For example, if we take the Lorenz system, we can construct the following master-slave system by replacing the state variable, y_m, of the master system in the slave system equations [55].

Master

$$\dot{x}_m = \sigma(y_m - x_m)$$
$$\dot{y}_m = x_m (\rho - z_m) - y_m \qquad (3)$$
$$\dot{z}_m = x_m y_m - \beta z_m$$

Slave

$$\dot{x}_s = \sigma(y_m - x_s)$$
$$\dot{z}_s = x_s y_m - \beta z_s \qquad (4)$$

Figure 6 shows the convergence of the time series, with different initial conditions, of the master and slave system for the state variable x.

Chaotic synchronization can be divided into the following categories considering the dynamics of master-slave systems, the synchronization error, and the coupling type [52, 56]:

- *Complete synchronization.* In this case, the synchronization error is defined as,

$$e(t) = (x_s(t), y_s(t), z_s(t)) - (x_m(t), y_m(t), z_m(t))$$

- *Generalized synchronization.* In this model, the state variables of one of the systems become functionally dependent on the other system's variables.

$$e(t) = (x_s(t), y_s(t), z_s(t)) - (\phi(x_m(t)), \phi(y_m(t)), \phi(z_m(t))) \tag{5}$$

- *Phase synchronization.* This method allows to synchronize two systems through their phases keeping the relation:

$$|m\phi_1(t) - n\phi_2(t)| < C \tag{6}$$

where m and n are integers, and the phases ϕ_1 and ϕ_2 of the systems are bounded by a constant C.
- *Lag synchronization.* In this method, the trajectories of the slave system converge to the trajectories of the master system at delayed time intervals $(t - \tau)$.

$$e(t) = (x_s(t), y_s(t), z_s(t)) - (x_m(t - \tau), y_m(t - \tau), z_m(t - \tau)) \tag{7}$$

- *Projective synchronization.* The state variables of one of the systems is scaled by a factor α with respect to the variables of the second system.

$$e(t) = (x_s(t), y_s(t), z_s(t)) - (\alpha_1 x_m(t), \alpha_2 y_m(t), \alpha_3 z_m(t)) \tag{8}$$

- *Anticipating synchronization.* Analogous to lag synchronization, the trajectories between the master system and the slave system follow the time relationship $(t + \tau)$.

$$e(t) = (x_s(t), y_s(t), z_s(t)) - (x_m(t + \tau), y_m(t + \tau), z_m(t + \tau)) \tag{9}$$

The synchronization models described above are based on different control strategies such as the Pecora-Carroll method, OGY, feedback control, active control, sliding mode control, non-linear control, and fuzzy control [14, 15, 52, 56].

2.4 Secure Communication Applications

Over the last decades, a large amount of research related to secure chaos-based communications has been conducted due to the natural relationship between chaos and security (cryptography), see Table 1. Many of these safe schemes are based on the synchronization of two chaotic systems. The master system is the transmitter, and the slave is the receiver. The chaotic system is used to hide the information to be transmitted. The states can be used to secure the communication in one of three chaos synchronization-based communications strategies: Chaotic Masking, in which the message is hidden to the output of the chaotic master systems and transmitted to the receiver, Parametric Modulation, where the message modulates the parameters of the chaotic carrier, and Chaotic Switching, in which a message signal is used to choose the carrier signal from different chaotic attractors [14, 16].

Table 1 Related chaotic and cryptographic properties [14]

Chaotic property	Cryptographic property	Description
Parameter and initial conditions sensitivity	Diffusion-key sensitivity	A small variation in input can cause large changes in output
Deterministic dynamics	Pseudo-randomness	Reproducible random behavior
Ergodicity	Confusion	The output has the same distribution for any input
Mixing property	Diffusion-plaintext sensitivity	A small local perturbation can cause large changes in the whole of space

Many chaos-based secure communication systems have been proposed based on the above three strategies or their variations with traditional encryption methods. However, some of them do not perform complete security analyses. Besides, security is ensured by the complexity of chaotic behavior. Therefore, four main aspects need to be addressed to evaluate the security levels of a secure scheme: dynamic, computational, design, and security [14–16].

- *Dynamical features* for analyzing and select a robust chaotic system, i.e., a chaotic system with high sensitivity to perturbations, significant chaotic range, large complex spectrum, and high "randomness".
- *Computational features* as computational efficiency, memory usage, and rounding errors.
- *Design features* such as encryption structure, encryption process, key space, and operations involved. It is always assumed that all details of the process and implementation are known to a third party.
- *Security features* to evaluate the resistance of the scheme against known attacks.

3 Methodology

The proposed secure scheme uses projective synchronization (PS) of two identical hyperchaotic systems in a master-slave configuration. The encryption process is composed of the discretization of the chaotic system and modulo operation as encryption transformation. This scheme is a symmetrical stream cipher like the "One-time pad" model, where chaotic sequences are like random noise and never reuse the secret key. We guarantee viability and robustness through dynamic, statistical, and security analysis.

3.1 Chaotic Sequences

The chaotic sequence is generated using the Qi hyperchaotic system, which has two large positive Lyapunov exponents with a wide range of chaotic parameters. In addition, this system presents a high degree of disorder in its trajectories and high-magnitude frequency bandwidths. These properties make this hyperchaotic system promising for security applications [57]. The Qi system is written as:

$$
\begin{aligned}
\dot{x}_1(t) &= S_{p_1}(x_2 - x_1) + S_{p_5}(x_2 x_3) - S_{p_7} x_4 \\
\dot{x}_2(t) &= S_{p_3} x_1 - S_{p_4} x_2 - x_1 x_3 \\
\dot{x}_3(t) &= x_1 x_2 - S_{p_2} x_3 \\
\dot{x}_4(t) &= S_{p_8} x_1 + S_{p_6}(x_2 x_3)
\end{aligned}
\tag{10}
$$

where x_i are state variables, and $S_{pj} \in \mathbb{R}$ the system parameters.

To generate the chaotic sequence, we use a numerical integration method based on trigonometric polynomials to solve the system of the differential equations [58]. The values of the discretized trajectories of the chaotic system are the elements of the chaotic sequence S_c given as:

$$
S_c = [\{x_4\}, \{x_3\}, \{x_2\}, \{x_1\}]
\tag{11}
$$

where $\{x_i\} = \{x_i(t_0), x_i(t_1), \ldots, x_i(t_n)\}$, and $i = 1, 2, 3, 4$.

We discard the first $0 < n_0 \leq N$ elements of $\{x_i\}$ in the sequence to achieve the expected synchronization error and guarantee the desirable convergence.

3.2 Synchronization Strategy

In this chapter, we use projective synchronization to synchronize trajectories of hyperchaotic systems. For this purpose, we use Lyapunov stability theory and define an active control to satisfy the stability conditions and achieve the desired error. Details and definitions are given below.

Let $\dot{x} = f(x)$ and $\dot{y} = g(y) + u(x, y)$ be the master system and the slave system respectively, where $f = (f_1, \ldots, f_n)^T, g = (g_1, \ldots, g_n)^T, u = (u_1, \ldots, u_n)^T$ is the vector control, and $x = (x_1, \ldots, x_n)^T, y = (y_1, \ldots, y_n)^T$ are the state vectors. If exist a constant matrix $A = \mathrm{diag}(\alpha_1, \ldots, \alpha_n)$ where $\alpha_i \neq 0, i = 1, \ldots, n$ such that

$$
\lim_{t \to \infty} \|e(t)\|_1 = \lim_{t \to \infty} \|y(t) - Ax(t)\|_1 = 0
\tag{12}
$$

where $e = (e_1, \ldots, e_n)$, $\|x\|_1 = \sum_{i=1}^{n} |x_i|$. Then the synchronization is called projective, and α_i is the scaling factor [56].

In our case, the master-slave system is given by the following hyperchaotic systems respectively:

$$
\begin{aligned}
\dot{x}_1 &= S_{p_1}(x_2 - x_1) + S_{p_5}x_2x_3 - S_{p_7}x_4 \\
\dot{x}_2 &= S_{p_3}x_1 - S_{p_4}x_2 - x_1x_3 \\
\dot{x}_3 &= x_1x_2 - S_{p_2}x_3 \\
\dot{x}_4 &= S_{p_8}x_1 + S_{p_6}x_2x_3
\end{aligned}
\tag{13}
$$

and

$$
\begin{aligned}
\dot{y}_1 &= S_{p_1}(y_2 - y_1) + S_{p_5}y_2y_3 - S_{p_7}y_4 + u_1 \\
\dot{y}_2 &= S_{p_3}y_1 - S_{p_4}y_2 - y_1y_3 + u_2 \\
\dot{y}_3 &= y_1y_2 - S_{p_2}y_3 + u_3 \\
\dot{y}_4 &= S_{p_8}y_1 + S_{p_6}y_2y_3 + u_4
\end{aligned}
\tag{14}
$$

where

$$
u(x, y) = -g(y) + Af(x) - Ke
\tag{15}
$$

are the active controls, $K = \mathrm{diag}(k_1, k_2, k_3, k_4)$, $k_i > 0$, $i = 1, 2, 3, 4$ is a real constant called the gain.

From (12) the error derivative is given by:

$$
\begin{aligned}
\dot{e} &= \dot{y} - A\dot{x} \\
\dot{e} &= g(y) + u(x, y) - Af(x)
\end{aligned}
\tag{16}
$$

By substituting (15) into (16) we obtain the closed-loop error dynamic:

$$
\dot{e} = -Ke
\tag{17}
$$

The dynamic error equation (17) has an equilibrium point at the origin $e_p = (0, 0, 0, 0)$.

To prove that synchronization error satisfies (12), the following Lyapunov function is selected as:

$$
V(e) = \tfrac{1}{2}(e_1^2 + e_2^2 + e_3^2 + e_4^2)
\tag{18}
$$

From (17)

$$
\begin{aligned}
\dot{V}(e) &= e_1\dot{e}_1 + e_2\dot{e}_2 + e_3\dot{e}_3 + e_4\dot{e}_4 \\
&= -k_1e_1^2 - k_2e_2^2 - k_3e_3^2 - k_4e_4^2
\end{aligned}
\tag{19}
$$

Then, from (18) to (19) the following two conditions are satisfied:

$$V(e_p) = 0 \tag{20}$$

$$\dot{V}(e) < 0 \quad \text{for all } e \neq e_p \tag{21}$$

Hence, the error is asymptotically stable at the origin by the Lyapunov stability theory [52, 56], and the synchronization is achieved.

3.3 Encryption/Decryption Process

The encryption/decryption process uses chaotic sequences to encrypt/decrypt the message through the modulo operation. Figure 7 depicts the flowchart of the process. The general process in Algorithm 1, and Algorithm 2 involves the following steps:

1. At the transmitter side, we generate a secret key composed of the initial conditions (I_c), system parameters (S_p), and scaling factors (α) of the hyperchaotic master-slave system.
2. We use the secret key to initialize the master-slave system and generate the chaotic sequence, S_c, of length equal to $D + 4n_0$, equation (11). The initial message size is D, and n_0 represents the first discarded elements of the chaotic sequence, $n_0 = 30000$ elements. For example, if the message is an image of 256×256, the length of S_c will be $256 \times 256 + 4n_0$.
3. At the same time, we generate the control variable, u, to synchronize the hyperchaotic master-slave system.
4. From the chaotic sequence, we perform a domain space transformation (discretization process). This transformation converts the real values into integer elements

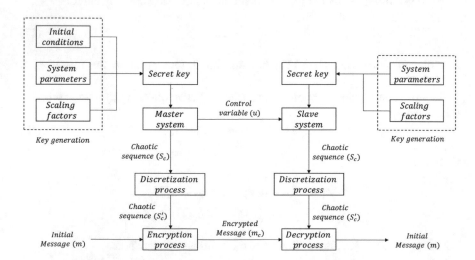

Fig. 7 Flowchart of the encryption scheme. The control variable is transmitted and used to generate the same set of values on the receiver side and perform the decryption process

through a decimal displacement of 7 units. For example, if the sequence value is 8.743176711, then the new component is 87431767. Thus, the domain transformation generates a new integer chaotic sequence S_c'.

5. We encrypt the initial message m to be transmitted through the modulo operation to obtain the encrypted message, m_c.

$$m_c = (m + S_c') \quad \text{mod } (b + 1) \tag{22}$$

where b is the maximum element value in the message, and S_c' the integer chaotic sequence.

6. We should send the encrypted message, m_c, and the control variable, u, through the public channel to the receiver.
7. Also, we must send the system parameters and scaling factors through a private channel to generate the decryption key. However, on the receiver side, the initial conditions of the systems could be any value and are not part of the secret key. It is due to the trajectories of the master-slave system converge for all initial conditions of the slave system by the synchronization process.
8. At the receiver side, the parties synchronize the state variables of the slave system with the master system using the control variable and generate the chaotic sequence S_c. Then, the receiver discretizes the chaotic sequence to obtain S_c' again.
9. Finally, to recover the original message, the receiver uses the chaotic sequence S_c' and the encrypted message m_c to obtain the original message m. Then, it is possible to recover the message m through the modulo operation by

$$m = (m_c - S_c') \quad \text{mod } (b + 1) \tag{23}$$

4   Results and Discussion

This section presents the performed experiments on the proposed algorithm and the results obtained in each test. The experiments are divided into three groups: dynamic analysis, statistical analysis, and security analysis. These analyses have a general objective to evaluate different properties showing the proposed algorithm's efficiency, robustness, and security levels.

The experiments and image encryption algorithms were performed in MATLAB© $R2021a$ academic. The test images were taken from the USC-SIPI database repository, and a personal computer of 2.40 GHz CORE $i5$ ×64 with 8 GB RAM and 1 TB hard disk capacity were used.

Algorithm 1 : Encryption Algorithm

Require: Initial message (m), initial conditions (I_c), system parameters (S_p), and scaling factors (α)

Ensure: $n_0 = 30000$

1: $D \leftarrow size(m)$ ▷ Obtain the image size

2:

3: **procedure** KEY GENERATION(I_c, S_p, α)

4:     $K \leftarrow (I_c, S_p, \alpha)$

5: **end procedure**

6:

7: **procedure** SEQUENCE GENERATION(K, D)

8:     $[\{x_1\}, \{x_2\}, \{x_3\}, \{x_4\}, u] \leftarrow ChaoticIntegration(K)$ ▷ Iterate the hyperchaotic system with the initial conditions, system paramters, and scaling factors for ($\frac{D}{4} + 4n_0$) values

9:     $S_c \leftarrow [\{x_4\}, \{x_3\}, \{x_2\}, \{x_1\}]$ ▷ Chaotic sequence

10: **end procedure**

11:

12: **procedure** DISCRETIZATION PROCESS(S_c)

13:     $S_c' \leftarrow fix(S_c * 10^7)$ ▷ The fix function extract the integer part of each element

14:     $S_c' \leftarrow reshape(S_c', D)$ ▷ Transform the sequence into a matrix

15: **end procedure**

16:

17: **procedure** ENCRYPTION PROCESS(S_c')

18:     $m_c \leftarrow (m + S_c') \mod (b + 1)$ ▷ Encrypted message

19: **end procedure**

Algorithm 2 : Decryption Algorithm

Require: Encrypted message (m_c), control variable (u), initial conditions (I_c), system parameters (S_p), and scaling factors (α)

Ensure: $n_0 = 30000$

1: $D \leftarrow size(m_c)$ ▷ Obtain the image size

2:

3: **procedure** KEY GENERATION(S_p, α)

4:     $K \leftarrow (S_p, \alpha)$

5: **end procedure**

6:

7: **procedure** SEQUENCE GENERATION(K, D, u, I_c)

8:     $[\{x_1\}, \{x_2\}, \{x_3\}, \{x_4\}] \leftarrow ChaoticIntegration(K, u, I_c)$ ▷ Iterate the hyperchaotic system with the secret key, control variable and initial conditions for ($\frac{D}{4} + 4n_0$) values

9:     $S_c \leftarrow [\{x_4\}, \{x_3\}, \{x_2\}, \{x_1\}]$ ▷ Chaotic sequence

10: **end procedure**

11:

12: **procedure** DISCRETIZATION PROCESS(S_c)

13:     $S_c' \leftarrow fix(S_c * 10^7)$ ▷ The fix function extract the integer part of each element

14:     $S_c' \leftarrow reshape(S_c', D)$ ▷ Transform the sequence into a matrix

15: **end procedure**

16:

17: **procedure** ENCRYPTION PROCESS(S_c')

18:     $m \leftarrow (m_c - S_c') \mod (b + 1)$ ▷ Initial message

19: **end procedure**

4.1 Dynamical Analysis

The core of the design of the chaotic-based secure schemes is the selection of the chaotic dynamical system. The system must have dynamical characteristics to satisfy the cryptographic requirements of confusion, diffusion, and random properties. Therefore, it is necessary to perform a dynamical analysis of the system to guarantee robustness against different attacks. We perform qualitative and quantitative analyzes of the master hyperchaotic system (13) to verify that satisfies the cryptographic requirements (see Sect. 2.4). These include spatial, temporal, spectral behavior, and dynamical properties. The initial conditions of the master system are $(0.5, 0.5, 0.5, 0.5)^T$, and the parameters are given in Table 2.

The numerical integration method was Gautschi [58]. This method has better precision than Forward Euler and is less expensive than a Runge-Kutta 4 [59]. The parameters of this numerical method used are step length, $\Delta t = 10^{-4}$, initial and final time, $t_i = 0, t_f = 3.2$. Besides, it requires declaring the system frequency, which was fixed as $w = 20$.

Figures 8, 9 and 10 depict the high complexity of the spatial evolution, time series, and frequency spectrum of the master hyperchaotic system. Figure 8 exhibits the "strange" attractor behavior and complex spatial dynamics of the master system, where it shows the density and disorder of the orbits. Figure 9 shows the complexity of the time series, where the system presents fast changes and strong randomness. In Fig. 10, the power spectral analysis reveals that the system has a complex and broad bandwidth, which implies that the signal change in a complicated way without exhibiting significant peaks or dominant frequencies. These features increase the robustness of the proposed scheme and make it difficult to extract and capture some information from the encrypted message and dynamic system.

Table 3 shows the dynamical properties of the hyperchaotic system of Qi [57]. It presents the number of system parameters, chaotic range, Kaplan-York dimension, Kolmogorov-Sinai entropy, bandwidth, equilibrium points, the minimum, and maximum value for the Lyapunov exponents, separation time. The domain of the parameters S_{p_1} and S_{p_2} of the chaotic system represents the chaotic range.

Table 2 System parameters

Parameter	Value
S_{p1}	58
S_{p2}	16
S_{p3}	49
S_{p4}	6
S_{p5}	15
S_{p6}	30
S_{p7}	35
S_{p8}	33

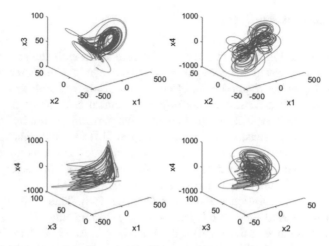

Fig. 8 Spatial behavior of the hyperchaotic Qi system (phase space)

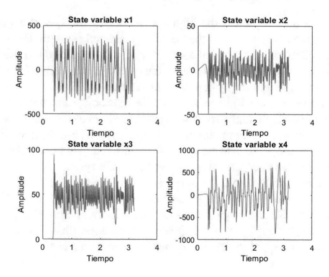

Fig. 9 Time series of the hyperchaotic Qi system for the four state variables (x_1, x_2, x_3, x_4)

The quantitative analysis shows the high dynamic complexity of the master hyperchaotic system. The system presents high sensitivity and unpredictability given by its Lyapunov exponents, which allows the expansion of the key space and provides evidence of the random noise-like behavior of the system. The dimension of Kapla-York enables us to observe the high structural complexity of the system, and the instability of its equilibrium points (see [57]) increases the robustness against dynamical attacks. Also, the system displays a wide chaotic range and parameter space, allowing to increase the key space. The spectral bandwidth is superior to traditional chaotic systems such as Lorenz, Rössler, Chen, and Lü, which have a bandwidth of approx-

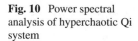

Fig. 10 Power spectral analysis of hyperchaotic Qi system

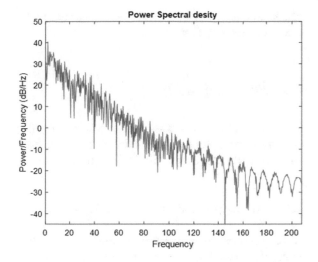

Table 3 Quantitative properties of the hyperchaotic Qi system [57]. The chaotic range represents the system parameter intervals where the system is hyperchaotic. Maximum *(Max)* and minimum *(Min)* Lyapunov exponent values are indicated

Quantitative properties	
Chaotic system	Hyperchaotic Qi system
Parameter number	8
Chaotic range	$S_{p_1} \in [24, 129]$, $S_{p_2} \in [2.1, 62.8]$
Kaplan-York dimension	3.198
Kolmogorov-Sinai entropy	0.31
Bandwidth	72–110 Hz
Equilibrium points (unstable)	5
Lyapunov exponents	$Min(7) - Max(12)$
Separation time	0.2 s

imately 4 Hz [14, 57]. This higher bandwidth allows hiding messages of higher frequency, $f > 4$ Hz, without showing distortions in the transmitted encrypted message. In addition, the system has a shorter separation time in the paths than other systems. This time makes it possible to demonstrate the sensitivity of the chaotic system and the rapid divergence of the trajectories, increasing the levels of diffusion of the encryption system.

The previous analysis includes a complete study of the dynamical properties of the hyperchaotic Qi system [57]. The analyzes allow us to know the characteristics of sensitivity, unpredictability, and randomness of the system. These characteristics give evidence of the properties of diffusion, confusion, and randomness of the whole system, increasing the proposed encryption algorithm's robustness and security levels. Therefore, this hyperchaotic system has desirable dynamical characteristics in a chaos-based encryption scheme.

4.2   Statistical Analysis

Statistical analysis allows studying the cryptographic system's confusion, diffusion, and randomness properties related to element distribution, correlation characteristics, and sensitivity properties. We describe these analyzes below.

4.2.1   Histogram and Entropy Analysis

A robust encryption process must have a uniform distribution of the encrypted message to increase the resistance to statistical attacks. Also, a uniform distribution is a measure of the randomness of the system. Here we use the histogram and entropy to characterize this property. Figure 11 shows the histogram of the original message and encrypted message. Appendix 6 presents histogram figures for other images in the database. Table 4 presents the chi-square test (χ^2) to quantify the uniformity of the histograms, where if we select a significance level of $\sigma = 0.05$, then $\chi^2(0.05) < 293.25$ to support the uniformity of the values in the histogram [60–62]. Table 5 gives the entropy values for different images.

Figure 11 shows the distribution of the elements in the original message and encrypted message. The figures depict the uniform values distribution of the encrypted

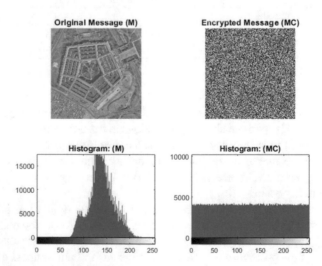

Fig. 11 Histogram values of the original message (Pentagon), and encrypted message. The corresponding histogram is shown below each figure

Table 4 χ^2 values for the encrypted images

Image	Histogram analysis encryped messages
	Critical value $\chi^2(0.05) < 293.25$
Pentagon	260.0641 ± 27.3717
Clock	248.9812 ± 29.8803
BlockGray	253.1177 ± 29.0756
CarToy	252.0918 ± 26.8719
Truck	254.9090 ± 21.6609
Walter	247.8820 ± 19.0133
Chemical Plant	247.6078 ± 21.2872
Cameraman	250.7214 ± 30.4450

Table 5 Entropy values for the original message, encrypted message, and chaotic sequence

Entropy values			
Image	Original message	Encrypted message	Chaotic sequence
Pentagon	6.7327 ± 0.0000	7.9998 ± 0.0000	7.9989 ± 0.0000
Clock	6.7057 ± 0.0000	7.8870 ± 0.0001	7.9972 ± 0.0001
BlockGray	4.3923 ± 0.0000	7.9993 ± 0.0001	7.9993 ± 0.0001
CarToy	6.2586 ± 0.0000	7.9993 ± 0.0001	7.9993 ± 0.0000
Truck	6.5632 ± 0.0000	7.9993 ± 0.0001	7.9993 ± 0.0001
Walter	7.2301 ± 0.0000	7.9972 ± 0.0002	7.9972 ± 0.0003
Chemical plant	7.3034 ± 0.0000	7.9973 ± 0.0002	7.9972 ± 0.0003
Cameraman	7.0097 ± 0.0000	7.9972 ± 0.0002	7.9972 ± 0.0001

information compared to the original signal, with a concentration of values in the histogram. Also, Table 4 presents the outcomes of the histogram analysis for different images, in which the values of χ^2 for all encrypted images is less than the critical value. In Table 5, we observe that the values for the encrypted message and chaotic sequence are closer to the maximum expected value, 8, for a gray-scale image. These analyses give evidence of the equiprobability of the values in the encrypted messages and chaotic sequences. In addition, this uniformity property limits the incidence of statistical attacks and provides evidence of the randomness characteristic of the system.

4.2.2 Correlation Analysis

A characteristic of images is strong correlations of pixels in different directions, where adjacent pixels have similar element values. This feature is undesirable in a

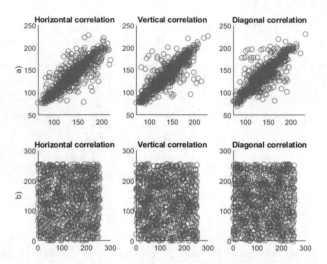

Fig. 12 Correlation figures of the original and encrypted message. Figure **a** horizontal, vertical, and diagonal correlation for the original message, and Figure **b** the correlation for encrypted image

robust encryption process. Hence, the low correlation between adjacent elements is a desirable property in image encryption. Figure 12 shows the vertical, horizontal, and diagonal correlation for the original (Pentagon) and encrypted signal. Appendix 6 presents correlation figures for other images in the database. Table 6 gives the correlation coefficients for different images.

Figure 12 illustrates the linear relation of the elements of the original message and encrypted message. Figure (a) shows a strong relationship between neighboring pixels of the original message, and Figure (b) shows the destruction of the linear relation between adjacent elements of the original message after the encryption process, which shows a low correlation and a homogeneous dispersion. These results are validated by evaluating 1000 messages and calculating the respective correlation coefficients before and after the encryption process, Table 6. The correlation values close to zero represent the destruction of the linear relationship, and elements around 1 show a strong relationship between neighboring message elements. These results reveal the randomness and confusion characteristics of the proposed encryption process and increased resistance against statistical attacks.

4.2.3 Differential Analysis

We measure the resistance to differential attacks through the NPCR and UACI tests, which analyzed the differences between two encrypted messages with slight perturbations on the original message or secret key. We define a significance level, $\sigma - level$, and associate a critical value N_σ^* and $U_\sigma^\pm$ to each test [20].

Table 6 Correlation coefficients of the original, encrypted message, and chaotic sequence

Correlation values

Image	Direction	Original message	Encrypted message	Chaotic sequence
Pentagon	H	0.8648 ± 0.0062	-0.0048 ± 0.0116	-0.0014 ± 0.0082
	V	0.8607 ± 0.0033	0.0027 ± 0.0155	-0.0006 ± 0.0149
	D	0.7936 ± 0.0066	0.0094 ± 0.0096	-0.0184 ± 0.0167
Clock	H	0.9567 ± 0.0032	0.0006 ± 0.0105	-0.0076 ± 0.0100
	V	0.9742 ± 0.0019	-0.0035 ± 0.0180	0.0042 ± 0.0090
	D	0.9381 ± 0.0039	-0.0027 ± 0.0098	0.0022 ± 0.0207
BlockGray	H	0.9961 ± 0.0015	0.0014 ± 0.0123	-0.0045 ± 0.0144
	V	0.9998 ± 0.0000	-0.0001 ± 0.0074	-0.0024 ± 0.0107
	D	0.9963 ± 0.0012	0.0020 ± 0.0108	0.0007 ± 0.0146
CarToy	H	0.9914 ± 0.0007	-0.0047 ± 0.0097	0.0006 ± 0.0139
	V	0.9771 ± 0.0014	-0.0025 ± 0.0156	0.0048 ± 0.0140
	D	0.9414 ± 0.0025	0.0000 ± 0.0203	-0.0013 ± 0.0111
Truck	H	0.9421 ± 0.0026	0.0080 ± 0.0127	0.0028 ± 0.0088
	V	0.9122 ± 0.0025	-0.0012 ± 0.0106	-0.0003 ± 0.0125
	D	0.8922 ± 0.0034	-0.0068 ± 0.0116	-0.0078 ± 0.0151
Walter	H	0.9878 ± 0.0010	-0.0016 ± 0.0135	-0.0018 ± 0.0207
	V	0.9911 ± 0.0007	0.0003 ± 0.0137	0.0022 ± 0.0198
	D	0.9758 ± 0.0018	0.0026 ± 0.0138	-0.0074 ± 0.0138
Chemical Plant	H	0.9510 ± 0.0022	0.0044 ± 0.0087	-0.0063 ± 0.0180
	V	0.9069 ± 0.0034	0.0018 ± 0.0214	0.0057 ± 0.0101
	D	0.8623 ± 0.0054	-0.0009 ± 0.0143	0.0017 ± 0.0140
Cameraman	H	0.9341 ± 0.0062	0.0046 ± 0.0117	0.0055 ± 0.0197
	V	0.9588 ± 0.0025	-0.0076 ± 0.0120	0.0016 ± 0.0115
	D	0.9066 ± 0.0052	-0.0043 ± 0.0136	0.0040 ± 0.0128

We perform a slight perturbation on a single secret key element to generate two different keys in this test. Then, the encryption process uses these two nearby keys with a variation of 10^{-6} to encrypt the same original message. Finally, we obtain two encrypted messages and compare their difference through UACI and NPCR tests. Tables 7 and 8 show the results for 1000 different secret keys. If the NPCR and UACI results are out of the acceptance interval, we reject the null hypothesis, and the encrypted message is not random-like. In this analysis, we define three significance levels $\sigma = [0.05, 0.01, 0.001]$, and we expect that 99% of the encrypted message passes each analysis.

Tables 7 and 8 provide the average value of applying the NPCR and UACI tests. Also, tables show the percentage of encrypted messages that are within the critical

Table 7 NPCR randomness test for the image encryption process

NPCR TEST		Theoretically NPCR critical value		
Image	NPCR value	$N^*_{0.05} = 99.569\%$	$N^*_{0.01} = 99.527\%$	$N^*_{0.001} = 99.534\%$
Pentagon	99.6091 ±0.0063	100% Pass	100% Pass	100% Pass
Clock	99.5949 ±0.0197	100% Pass	100% Pass	100% Pass
BlockGray	99.6128 ±0.0104	100% Pass	100% Pass	100% Pass
CarToy	99.6129 ±0.0108	100% Pass	100% Pass	100% Pass
Truck	99.6128 ±0.0104	100% Pass	100% Pass	100% Pass
Walter	99.5969 ±0.0188	100% Pass	100% Pass	100% Pass
Chemical Plant	99.5986 ±0.0187	100% Pass	100% Pass	100% Pass
Cameraman	99.5956 ±0.0208	100% Pass	100% Pass	100% Pass

Table 8 UACI randomness test for image encryption process

UACI TEST		Theoretically UACI critical value		
Image	UACI value	$U^-_{0.05} = 33.282\%$ $U^+_{0.05} = 33.645\%$	$U^-_{0.01} = 33.225\%$ $U^+_{0.01} = 33.702\%$	$U^-_{0.001} = 33.159\%$ $U^+_{0.001} = 33.768\%$
Pentagon	33.4592 ±0.0244	100% Pass	100% Pass	100% Pass
Clock	33.4619 ±0.0597	100% Pass	100% Pass	100% Pass
BlockGray	33.4565 ±0.0247	100% Pass	100% Pass	100% Pass
CarToy	33.4439 ±0.0341	100% Pass	100% Pass	100% Pass
Truck	33.4651 ±0.0335	100% Pass	100% Pass	100% Pass
Walter	33.4442 ±0.0954	100% Pass	100% Pass	100% Pass
Chemical Plant	33.4574 ±0.1134	100% Pass	100% Pass	100% Pass
Cameraman	33.4663 ±0.0847	100% Pass	100% Pass	100% Pass

values. 100% of the encrypted messages pass the analysis successfully. However, the acceptance or rejection depends on the selection of the significance level.

The experimental results above show the differences between pairs of encrypted messages when we use keys with variations of 10^{-6}, enabling characterize and study the relation between the elements in the encrypted information. Furthermore, it gives evidence of the random behavior of the encrypted message. These analyzes support the confusion and diffusion properties of the proposed chaotic scheme. Therefore, the cryptosystem reduces the effectiveness of a differential attack.

4.3 Security Analysis

We need to evaluate the cryptosystems with some security analysis. Generally, this analysis cannot include all possible tests and attacks. Instead, it should cover some of the best-known cryptographic tests shown below.

4.3.1 Key Definition

We define the key space through its components' sensitivity, ranges, and restrictions (initial conditions, system parameters, and scaling factors).

- Eight initial conditions, I_c.
- Two system parameters, $S_p = (S_{p_1}, S_{p_2})$.
- Four scaling factors, $\alpha = (\alpha_1, \alpha_2, \alpha_3, \alpha_4)$.

Then, the secret key is composed of $K = (I_c, S_p, \alpha)$, and the key space is:

$$
|K| = |I_c|^8 \, |S_p|^2 \, |\alpha|^4
$$

$$
|K| = \left| \frac{n_1}{\left(10^{-12}\right)^8} \right| \left| \frac{n_2 n_3}{\left(10^{-12}\right)^2} \right| \left| \frac{n_4}{\left(10^{-12}\right)^4} \right| \tag{24}
$$

$$
|K| = n_1 n_2 n_3 n_4 \times 10^{96+24+48}
$$

$$
|K| = n_5 \times 10^{168}, \quad n_5 = n_1 n_2 n_3 n_4
$$

where $n_1, n_2, n_3, n_4 \in \mathbb{R}$. Initial conditions, system parameters, and scaling factors sensitivity are all equal to 10^{-12}.

The ranges and restrictions of the secret key components are defined by:

- Initial conditions: $I_c \in \mathbb{R}$
- System parameters: $S_{p_1} \in [24, 129]$, $S_{p_2} \in [2.1, 62.8]$
- Scaling factors: $\alpha \in \mathbb{R} \setminus [-1, 1]$

The secret key was defined with the following fixed parameters:

$$(S_{p_3}, S_{p_4}, S_{p_5}, S_{p_6}, S_{p_7}, S_{p_8}) = (49, 6, 15, 30, 35, 33)$$

to preserve the chaotic behavior of the system.

The key space of the proposed scheme is $K \approx 1 \times 10^{168}$, given by the sensitivity of the system parameters, which decreases the probability of a successful brute-force attack, see Table 10. This key space exceeds the minimum expected value of $2^{256} \approx 1.1 \times 10^{77}$.

4.3.2   Avalanche Effect

Avalanche effects enable the assessment of the sensitivity to small perturbations of the cryptographic scheme. At least 90% of the encrypted message elements should change before a slight variation on the secret key. Table 9 shows the variation percentage between two encrypted messages using secret keys with differences of 1×10^{-7} as follows:

Initial key (k_1):
   Initial conditions (I_c): (3.0, 0.5, 0.5, 0.6, 0.1, 0.1, 0.1, 0.1)
   System parameters (S_{p_1}, S_{p_2}): (56, 16)
   Scaling factors (α_i): (−3, 2, 3, 5)

Keys with slight variation
   Key (k_2):
   Initial conditions (I_c): (3.0, 0.5 + 1 × 10^{-7}, 0.5, 0.6, 0.1, 0.1, 0.1, 0.1)
   System parameters (S_{p_1}, S_{p_2}): (56, 16)
   Scaling factors (α_i): (−3, 2, 3, 5)

   Key (k_3):
   Initial conditions (I_c): (3.0, 0.5, 0.5, 0.6, 0.1, 0.1, 0.1, 0.1)
   System parameters (S_{p_1}, S_{p_2}): (56 + 1 × 10^{-7}, 16)
   Scaling factors (α_i): (−3, 2, 3, 5)

Table 9 indicates the results of the avalanche effect, in which we observe that a slight perturbation in the secret key produces a different encrypted message and chaotic sequence.

The perturbation produces a variation of more than 99% in the encrypted message and chaotic sequences elements, which exceeds the minimum variation value expected of 90%. This degree of sensitivity and variation improves and increases the diffusion property.

Table 9 Variation percentage for two different encrypted messages with different secret keys. Column 2 represents the percentage of variations between an encrypted message generated with the key 1 and a second message encrypted with the key 2. Column 3 the percentage of variations for the key 1 and the key 3

Avalanche test	Variation percentage (%) Encrypted messages	
Image		
Pentagon	99.6156	99.5926
Clock	99.6262	99.9900
BlockGray	99.6319	99.5975
CarToy	99.6319	99.5975
Truck	99.6319	99.5975
Walter	99.6262	99.9900
Chemical plant	99.6262	99.9900
Cameraman	99.6262	99.9900

4.3.3 Decryption Analysis

Decryption analysis enables the assessment of the sensitivity of the proposed scheme to variations in the secret key; that is, the slightest changes in the key components make it impossible for a third party to decrypt the message. This characteristic improves the robustness of the scheme to attacks using similar keys.

We use two keys with a variation of 1×10^{-12} to encrypt and decrypt the messages. First, the transmitter uses *key 1* to encrypt the original message. Then, an attacker uses a second key to decrypt the message. This second key is the encryption key (*key 1*) with a variation in one of its components. Figure 13 shows the original message (Clock), the encrypted message using *key 1*, and the recovered (decrypted) message using *key 2*.

Encryption key (k_1):
   Initial conditions (I_c): $(4.0, 0.5, 0.5, 0.6, 0.2, 0.1, 0.1, 0.1)$
   System parameters (S_{p_1}, S_{p_2}): $(26, 16)$
   Scaling factors (α_i): $(-5, 8, 3, 6)$

Decryption key (k_2):
   Initial conditions (I_c): $(4.0 + 1 \times 10^{-12}, 0.5, 0.5, 0.6, 0.2, 0.1, 0.1, 0.1)$
   System parameters (S_{p_1}, S_{p_2}): $(26, 16)$
   Scaling factors (α_i): $(-5, 8, 3, 6)$

Figure 13 illustrates the impossibility of performing a decryption attack using keys with slight variations, where a variation of 1×10^{-12} on the secret key is enough to prevent the recovery of the original message. This characteristic limits the incidence of this type of attack and gives evidence of the proposed chaotic scheme's sensitivity.

Original message (M)   Encrypted message (MC) Dencrypted message (MD)

a)

Fig. 13 Decryption attack with a secret key, k_2, with slightest differences in relation to the encryption key, k_1. The original message is encrypted with key 1 and decrypted with key 2

4.3.4 Distortion Analysis

The distortion attack applies different disturbances and modifications to the encrypted message such that it is impossible to recover all the original information. We analyze this attack using Gaussian noise and salt-and-pepper noise at 30, 50, and 60% of noisy density. Figures 14, 15, 16, and 17 present the original, encrypted, and decrypted messages for different kinds of noise.

Figures 14, 15, 16, and 17 show a distortion attack performed on the proposed algorithm using Gaussian and Salt-and-pepper noise in different percentages. This distortion analysis studies the resistance of the scheme to perturbations and modifications in the encrypted message. Figures 14 and 15 displays the encrypted message with different Gaussian noise percentages and their corresponding decrypted mes-

Original message (M)   Encrypted message (MC) Decrypted message (MD)

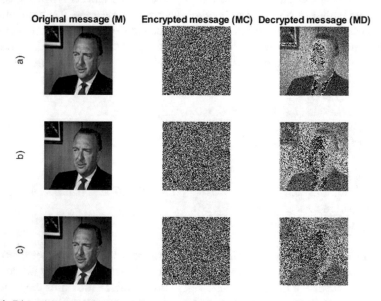

a)

b)

c)

Fig. 14 Distortion attack based on Gaussian noise in walter image. **a** Gaussian noise at 30%, **b** Gaussian noise at 50%, and **c** Gaussian noise at 60%

Fig. 15 Distortion attack based on Gaussian noise in walter chemical plant image. **a** Gaussian noise at 30%, **b** at 50%, and **c** at 60%

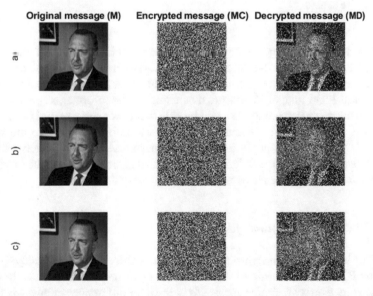

Fig. 16 Distortion attack based on salt-and-pepper noise in Walter image. Original message, encrypted noise message, and recovered message. Figure **a** shows the Sal-and-pepper noise at 30%, **b** at 50%, and **c** at 60%

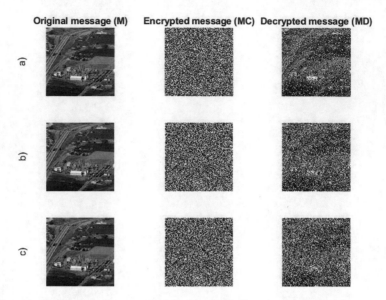

Fig. 17 Distortion attack based on salt-and-pepper noise in chemical plant image. Original message, encrypted noise message, and recovered message. Figure **a** shows the Sal-and-pepper noise at 30%, **b** at 50%, and **c** at 60%

sage. The decrypted image in Fig. 14 preserves more information of the original message than Fig. 15 with the same noise. This variation is due to the distribution of pixel elements in the Walter image, where we can observe a higher contrast than in the Chemical Plant image. Therefore, depending on the information content of the image and the noise percentage, we will be able to recover some partial information. In addition, Figs. 16 and 17 reveal that only partial information of the transmitted message is lost when we apply a salt-and-pepper noise. We can observe that the encryption method is more efficient with Salt-and-pepper. Therefore, the encryption scheme supports noisy scenarios with certain recovery limitations.

4.3.5 Differential Cryptanalysis

One of the best-known Chosen-plaintext attacks is Differential cryptanalysis introduced by Biham and Shamir [63]. This attack analyzes the differences in cipher message pairs from two similar messages with slight differences or encrypted messages produced by similar secret keys. This attack aims to find the most probable key and reduce the search space compared to a brute force attack. We apply the differential analysis to image encryption using the NPCR and UACI tests. See Tables 7 and 8.

4.3.6 Return Map Analysis

The return map attacks explore the fluctuation of the return maps of the transmitted signal to perform a partial reconstruction of the chaotic dynamics. A direct return map analysis uses the maps given by plotting the state variables $x_i(n)$ and $x_i(n + 1)$ to analyze the evolution of the encrypted signal on those maps. Through this analysis, we could extract the original message or secret key. A return map attack is a Ciphertext-only attack. Figure 18 present the original message (Truck) and the encrypted message. Figure 19 shows the return map, $m_c(n)$ versus $m_c(n + 1)$, of the transmitted signal (encrypted message) for 5000 values.

Figure 19 presents the return map of the transmitted signal. The map shows the complexity of the encrypted message, which looks like a random map distribution without any pattern that can reveal the system dynamics.

Fig. 18 Original message and encrypted message of the return map

Fig. 19 Return map of the transmitted signal (encrypted message). The map is plotting for the state variable $m_c(n)$ versus $m_c(n + 1)$

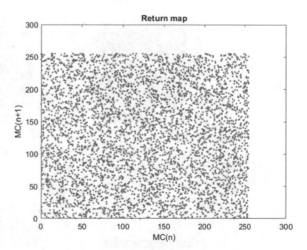

4.3.7   Known/Chosen Plaintext Attack Analysis

Known plaintext and chosen-plaintext attacks are the most common attacks for image cryptosystems. Therefore, we support the resistance to these attacks by analyzing the encryption effect of the all-black and all-white images of the encryption scheme. We expect a uniform distribution of encrypted messages from both images [2, 4]. Figures 20 and 21 present the histogram of the original and encrypted messages.

From Figs. 20 and 21, we observe a uniform distribution of the histograms, which means that we cannot obtain any information from the original message. Therefore, an attacker cannot recover valuable information of the secret key.

4.3.8   Parameter Estimation

Secret parameter estimation is a critical problem that affects many of the proposed chaotic cryptographic schemes. This cryptanalysis takes advantage of low sensitivity to parameters mismatch of the system to estimate the secret key and break the cryptographic scheme. Table 10 shows a parameter mismatch (sensitivity analysis of key components) based on encryption analysis using different keys with slight differences in the encryption and decryption process.

Parameters system sensitivity allows studying the robustness of the proposed scheme to different attacks whose objective is to break the cryptosystem using approximate keys to extract partial information about the original message or the chaotic dynamics. The high sensitivity of the proposed scheme, Table 10, reduces and limits the effectivity of different attacks as generalized and adaptive synchro-

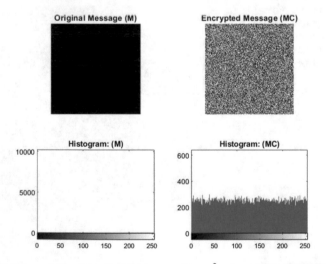

Fig. 20 Histogram analysis of the all-white image with a χ^2 value equal to 250.6250

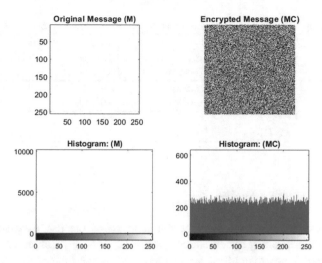

Fig. 21 Histogram analysis of the all-white image with a χ^2 value equal to 269.1719

Table 10 Sensitivity to small perturbations of the secret key components. The sensitivity is related to the minimum variation on a single system parameter, initial condition or a scaling factor

Secret key element	Sensitivity
System parameters	10^{-12}
Initial conditions	10^{-12}
Scaling factors	10^{-12}

nization, bit-error-rate estimation, and brute force attacks, which take advantage of limitations in sensitivity and parameters mismatch.

4.3.9 Time Analysis

We perform a time analysis for a size 256×256 grayscale image. Table 11 shows the encryption and decryption time of the proposed scheme. The encryption and decryption time includes generating chaotic sequences and the encryption/decryption process through modulo operation.

Table 11 presents a time analysis of the proposed scheme for a 256×256 and 512×512 image. The encryption/decryption times include generating chaotic sequence through Gautschi numerical method and the encryption/decryption process through modulo operation. As seen in the table, the encryption and decryption process requires little time compared to the generation of chaotic sequences. It is due to the use of numerical integration methods and the number of elements generated.

Table 11 Encryption and decryption time for an 256×256 and 512×512 image size

Image size	Encryptiontime (s)		Decryptiontime (s)	
	Chaotic sequence	Encryption process	Chaotic sequence	Decryption process
256×256	0.7574 ±0.0407	0.0127 ±0.0012	0.7574 ±0.0407	0.0091 ±0.0000
512×512	1.5647 ±0.1167	0.307 ±0.0016	1.5647 ±0.1167	0.253 ±0.0015

4.3.10 Comparative Analysis

In this section, we present a comparative analysis of the proposed scheme with some related works using hyperchaotic systems [2, 3, 6, 39]. Tables 12, 13, 14, and 15 depict a comparative analysis for different gray-scale images in terms of mean values for information entropy, correlation, histogram analysis (χ^2 Test), NPCR and UACI. The values provided in Tables 12, 13, 14, and 15 show the effectiveness and

Table 12 Comparative analysis for the proposed scheme and some related approaches for "Baboon" image

Image	Entropy value	Correlation			χ^2 Test	NPCR value	UACI value
Baboon		H	V	D			
Proposed	7.997	−0.021	0.013	−0.017	262.602	99.609	33.528
Xu et al. [2]	7.999	0.008	0.007	0.001	–	–	–
Li et al. [3]	7.999	0.005	0.002	0.002	–	99.621	33.431
Zhao and Ren [6]	–	–	–	–	–	–	–
Li and Zhang [39]	7.998	−0.002	0.001	0.005	–	99.620	33.431

Table 13 Comparative analysis for the proposed scheme and some related approaches for "Peppers" image

Image	Entropy value	Correlation			χ^2 Test	NPCR value	UACI value
Peppers		H	V	D			
Proposed	7.999	0.010	−0.001	0.017	223.045	99.602	33.452
Xu et al. [2]	–	–	–	–	–	–	–
Li et al. [3]	7.999	−0.005	0.006	0.002	–	99.603	33.503
Zhao and Ren [6]	–	–	–	–	–	–	–
Li and Zhang [39]	7.999	−0.001	−0.002	0.001	–	99.609	33.441

Table 14 Comparative analysis for the proposed scheme and some related approaches for "Cameraman" image

Image	Entropy value	Correlation			χ^2 Test	NPCR value	UACI value
Cameraman		H	V	D			
Proposed	7.999	0.010	−0.001	0.017	223.045	99.602	33.452
Xu et al. [2]	7.994	0.008	0.007	0.001	–	–	–
Li et al. [3]	–	–	–	–	–	–	–
Zhao and Ren [6]	7.997	0.001	0.003	0.000	–	99.611	33.468
Li and Zhang [39]	7.998	−0.001	0.001	0.005	–	99.620	33.431

Table 15 Comparative analysis for the proposed scheme and some related approaches for "Boat" image

Image	Entropy value	Correlation			χ^2 Test	NPCR value	UACI value
Boat		H	V	D			
Proposed	7.999	0.003	0.010	0.008	263.523	99.644	33.460
[2]	7.999	−0.002	−0.004	−0.007	–	–	–
[3]	7.999	−0.008	−0.001	−0.003	–	99.607	33.488
[6]	7.999	0.000	0.002	−0.003	–	99.610	33.466
[39]	7.999	−0.001	0.001	−0.001	–	99.609	33.468

similarities of the proposed encryption scheme with other works. The histogram analysis is not performed by any of the approaches analyzed.

5 Conclusions

The projective synchronization problem for the hyperchaotic Qi system and its application to secure communication was addressed. First, active control is designed to achieve chaotic projective synchronization between two identical systems. Then, the synchronization strategy was used to propose a secure communication scheme based on the hyperchaotic Qi system for image protection. Also, the proposed scheme was evaluated by dynamic, statistical, and security analysis. Finally, we gave evidence of the robustness and applicability of the Qi hyperchaotic system to secure communications or cryptography.

The dynamical analysis showed the characteristics and complex behavior of the chaotic system. It presented high sensitivity, wide chaotic range, high structural complexity, and high randomness, features that are suitable for encryption.

Statistical studies provided insight into the relationship, sensitivity, and element distribution of the signals involved in the process. It was showed that the secure chaotic scheme destroys the internal relationship between the elements of the original message. In addition, the high sensitivity to key variation was also observed, making it difficult to recover the original message in a decryption process. Besides, the randomness properties of the scheme are supported by different tests.

The security analysis enabled us to exhibit the robustness of the encryption scheme against different known attacks, such as distortion, differential, decryption, return mapping, brute force, parameter estimation, adaptive and generalized synchronization, and statistical attacks. Thus, the above analysis provides evidence of the security levels and feasibility of the proposed chaos-based secure scheme.

On the other hand, the analysis and experiments showed that projective synchronization of hyperchaotic systems with encryption strategies improves the security of secure communication schemes. It is produced by the complexity and sensitivity of hyperchaotic systems, the selection of encryption strategies such as One-time-Pad, and the robustness of projective synchronization. Also, the sensitivity of the scaling factors involved in the synchronization process increased the key space and improved the security levels of the scheme.

Finally, the comparative analysis showed the similarities and robustness of the proposed encryption scheme with some existing hyperchaotic proposals, where any of the analyzed schemes did not perform the histogram analysis (χ^2 test). Moreover, the analyzed works focus on statistical tests as robustness and security measures, which are necessary but not sufficient to guarantee the whole system's security.

6 Future Work

We will consider the use of different synchronization strategies, such as fuzzy synchronization. In addition, the use of more complex hyperchaotic systems, such as fractional ones, will be discussed.

Due to the lack of standards in chaotic cryptography, the performed tests only provide evidence of the robustness and feasibility of the proposed scheme. Therefore, the development of more accurate testing and analysis tools is necessary.

The main disadvantage of image encryption is the increased encryption time due to the amount of data. However, this processing time can be decreased by using parallel computing.

Acknowledgements The first author is thankful to Consejo Nacional de Ciencia y Tecnología (CONACYT) for the scholarship 937653.

Appendix

The appendix contains histogram and correlation analysis for different images from the USC-SIPI database (Figs. 22, 23, 24, 25, 26, 27, 28, 29, 30, 31, 32, 33, 34 and 35).

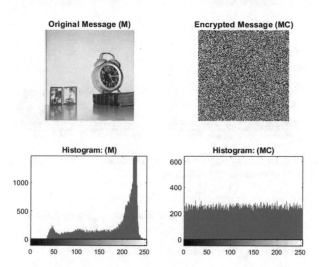

Fig. 22 Histogram values of the original message (Clock), and encrypted message. The corresponding histogram is shown below each figure

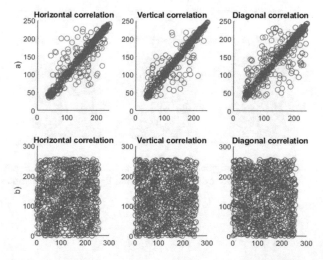

Fig. 23 Correlation figures of the original (Clock) and encrypted message. Figure **a** horizontal, vertical, and diagonal correlation for original message, and figure **b** the correlation for encrypted image

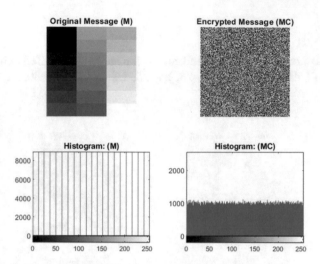

Fig. 24 Histogram values of the original message (BlockGray), and encrypted message. The corresponding histogram is shown below each figure

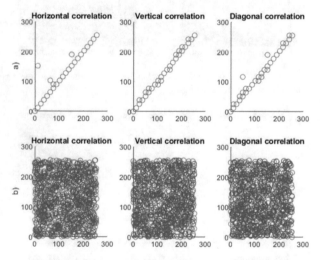

Fig. 25 Correlation figures of the original (BlockGray) and encrypted message. Figure **a** horizontal, vertical, and diagonal correlation for original message, and figure **b** the correlation for encrypted image

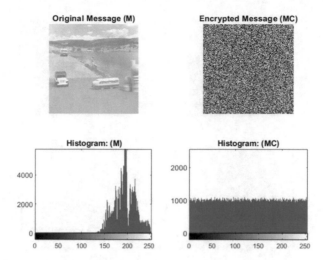

Fig. 26 Histogram values of the original message (CarToy), and encrypted message. The corresponding histogram is shown below each figure

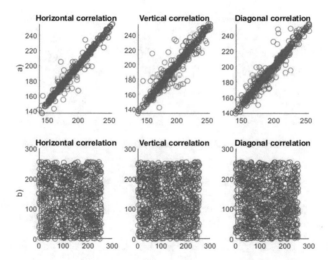

Fig. 27 Correlation figures of the original (CarToy) and encrypted message. Figure **a** horizontal, vertical, and diagonal correlation for original message, and figure **b** the correlation for encrypted image

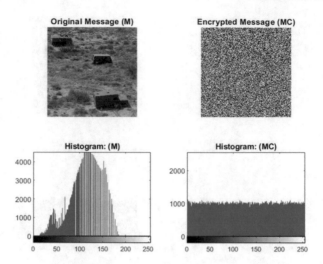

Fig. 28 Histogram values of the original message (Truck), and encrypted message. The corresponding histogram is shown below each figure

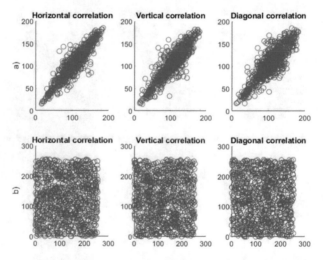

Fig. 29 Correlation figures of the original (Truck) and encrypted message. Figure **a** horizontal, vertical, and diagonal correlation for original message, and figure **b** the correlation for encrypted image

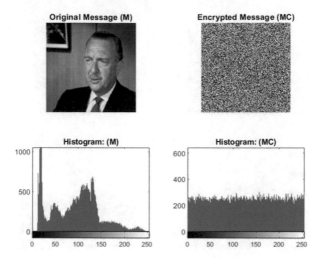

Fig. 30 Histogram values of the original message (Walter), and encrypted message. The corresponding histogram is shown below each figure

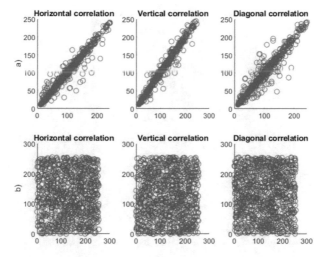

Fig. 31 Correlation figures of the original (Walter) and encrypted message. Figure **a** horizontal, vertical, and diagonal correlation for original message, and figure **b** the correlation for encrypted image

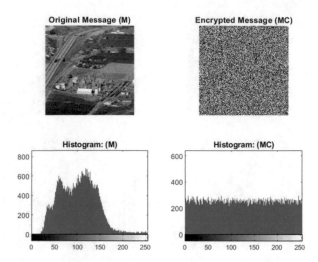

Fig. 32 Histogram values of the original message (chemical plant), and encrypted message. The corresponding histogram is shown below each figure

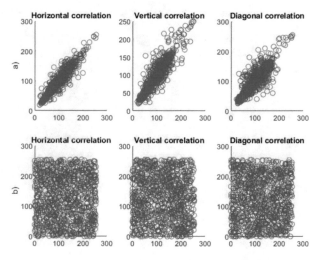

Fig. 33 Correlation figures of the original (chemical plant) and encrypted message. Figure **a** horizontal, vertical, and diagonal correlation for original message, and figure **b** the correlation for encrypted image

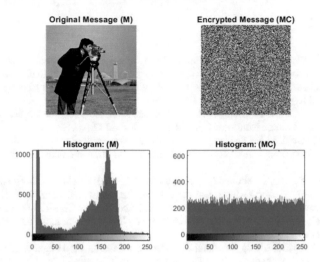

Fig. 34 Histogram values of the original message (cameraman), and encrypted message. The corresponding histogram is shown below each figure

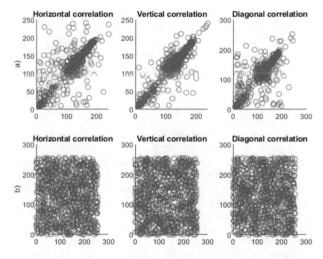

Fig. 35 Correlation figures of the original (cameraman) and encrypted message. Figure **a** horizontal, vertical, and diagonal correlation for original message, and figure **b** the correlation for encrypted image

References

1. Zhang, X., Wang, L., Wang, Y., Niu, Y., Li, Y.: An image encryption algorithm based on hyperchaotic system and variable-step josephus problem. Int. J. Opt. **2020** (2020)
2. Xu, Q., Sun, K., Cao, C., Zhu, C.: A fast image encryption algorithm based on compressive sensing and hyperchaotic map. Opt. Lasers Eng. **121**, 203–214 (2019)
3. Li, Z., Peng, C., Li, L., Zhu, X.: A novel plaintext-related image encryption scheme using hyper-chaotic system. Nonlinear Dyn. **94**(2), 1319–1333 (2018)
4. Ahmad, M., Doja, M.N., Beg, M.M.S.: Security analysis and enhancements of an image cryptosystem based on hyperchaotic system. J. King Saud Univ.-Comput. Inf. Sci. **33**(1), 77–85 (2021)
5. Tsafack, N., Sankar, S., Abd-El-Atty, B., Kengne, J. Jithin,K., Belazi, A., Mehmood, I., Bashir, A.K., Song, O.-Y., Abd El-Latif, A.A.: A new chaotic map with dynamic analysis and encryption application in internet of health things. IEEE Access **8**, 137731–137744 (2020)
6. Zhao, C.-F., Ren, H.-P.: Image encryption based on hyper-chaotic multi-attractors. Nonlinear Dyn. **100**(1), 679–698 (2020)
7. Chen, J., Wong, K., Cheng, L., Shuai, J.: A secure communication scheme based on the phase synchronization of chaotic systems. Chaos: Interdiscip. J. Nonlinear Sci. **13**(2), 508–514 (2003)
8. Kocarev, L., Halle, K., Eckert, K., Chua, L.O., Parlitz, U.: Experimental demonstration of secure communications via chaotic synchronization. Int. J. Bifurc. Chaos **2**(03), 709–713 (1992)
9. Memon, Q.A.: Synchronized choas for network security. Comput. Commun. **26**(6), 498–505 (2003)
10. Wu, Z., Zhang, X., Zhong, X.: Generalized chaos synchronization circuit simulation and asymmetric image encryption. IEEE Access **7**, 37989–38008 (2019)
11. Parlitz, U., Chua, L.O., Kocarev, L., Halle, K., Shang, A.: Transmission of digital signals by chaotic synchronization. Int. J. Bifurc. Chaos **2**(04), 973–977 (1992)
12. Pellicer-Lostao, C., Lopez-Ruiz, R.: Notions of chaotic cryptography: sketch of a chaos based cryptosystem (2012). arXiv:1203.4134
13. Alvarez, G., Li, S., Montoya, F., Pastor, G., Romera, M.: Breaking projective chaos synchronization secure communication using filtering and generalized synchronization. Chaos Solitons Fractals **24**(3), 775–783 (2005)
14. Alvarez, G., Li, S.: Some basic cryptographic requirements for chaos-based cryptosystems. Int. J. Bifurc. Chaos **16**(08), 2129–2151 (2006)
15. Alvarez, G., Amigó, J.M., Arroyo, D., Li, S.: Lessons learnt from the cryptanalysis of chaos-based ciphers. In: Chaos-Based Cryptography, pp. 257–295. Springer (2011)
16. Bendoukha, S., Abdelmalek, S., Ouannas, A.: Secure communication systems based on the synchronization of chaotic systems. In: Mathematics Applied to Engineering, Modelling, and Social Issues, pp. 281–311. Springer (2019)
17. Özkaynak, F.: Brief review on application of nonlinear dynamics in image encryption. Nonlinear Dyn. **92**(2), 305–313 (2018)
18. Dachselt, F., Schwarz, W.: Chaos and cryptography. IEEE Trans. Circ. Syst. I: Fundam. Theory Appl. **48**(12), 1498–1509 (2001)
19. Li, S., Alvarez, G., Li, Z., Halang, W.A.: Analog chaos-based secure communications and cryptanalysis: a brief survey (2007). arXiv:0710.5455
20. Wu, Y., Noonan, J.P., Agaian, S., et al.: NPCR and UACI randomness tests for image encryption. Cyber J.: Multidiscip. J. Sci. Technol. J. Sel. Areas Telecommun. (JSAT) **1**(2), 31–38 (2011)
21. Alvarez, G., Li, S.: Breaking network security based on synchronized chaos. Comput. Commun. **27**(16), 1679–1681 (2004)
22. Sambas, A., Vaidyanathan, S., Tlelo-Cuautle, E., Abd-El-Atty, B., Abd El-Latif, A.A., Guillén-Fernández, O., Hidayat, Y., Gundara, G., et al.: A 3-D multi-stable system with a peanut-shaped equilibrium curve: circuit design, FPGA realization, and an application to image encryption. IEEE Access **8**, 137116–137132 (2020)

23. Vaidyanathan, S., Sambas, A., Abd-El-Atty, B., Abd El-Latif, A.A., Tlelo-Cuautle, E., Guillén-Fernández, O., Mamat, M., Mohamed, M.A., Alçin, M., Tuna, M., et al.: A 5-D multi-stable hyperchaotic two-disk dynamo system with no equilibrium point: circuit design, fpga realization and applications to trngs and image encryption. IEEE Access (2021)
24. Mostafaee, J., Mobayen, S., Vaseghi, B., Vahedi, M., Fekih, A.: Complex dynamical behaviors of a novel exponential hyper-chaotic system and its application in fast synchronization and color image encryption. Sci. Prog. **104**(1), 00368504211003388 (2021)
25. Javan, A.A.K., Jafari, M., Shoeibi, A., Zare, A., Khodatars, M., Ghassemi, N., Alizadehsani, R., Gorriz, J.M.: Medical images encryption based on adaptive-robust multi-mode synchronization of chen hyper-chaotic systems. Sensors **21**(11), 3925 (2021)
26. Muthukumar, P., Balasubramaniam, P., Ratnavelu, K.: A novel cascade encryption algorithm for digital images based on anti-synchronized fractional order dynamical systems. Multimed. Tools Appl. **76**(22), 23517–23538 (2017)
27. Di, X., Li, J., Qi, H., Cong, L., Yang, H.: A semi-symmetric image encryption scheme based on the function projective synchronization of two hyperchaotic systems. PloS one **12**(9), e0184586 (2017)
28. Zhang, F., Liu, J., Wang, Z., Jiang, C.: N-systems function projective combination synchronization–A review of real and complex continuous time chaos synchronization. IEEE Access **7**, 179320–179338 (2019)
29. Alvarez, G., Montoya, F., Romera, M., Pastor, G.: Breaking two secure communication systems based on chaotic masking. IEEE Trans. Circ. Syst. II: Express Briefs **51**(10), 505–506 (2004)
30. Yang, T., Yang, L.-B., Yang, C.-M.: Breaking chaotic secure communication using a spectrogram. Phys. Lett. A **247**(1–2), 105–111 (1998)
31. Singh, S., Ahmad, M., Malik, D.: Breaking an image encryption scheme based on chaotic synchronization phenomenon. In: 2016 Ninth International Conference on Contemporary Computing (IC3), pp. 1–4. IEEE (2016)
32. Ahmad, M., Aijaz, A., Ansari, S., Siddiqui, M.M., Masood, S.: Cryptanalysis of image cryptosystem using synchronized 1D lorenz stenflo hyperchaotic systems. In: Information and Decision Sciences, pp. 367–376. Springer (2018)
33. Li, C., Lo, K.-T.: Optimal quantitative cryptanalysis of permutation-only multimedia ciphers against plaintext attacks. Signal Process. **91**(4), 949–954 (2011)
34. Li, C.: Cracking a hierarchical chaotic image encryption algorithm based on permutation. Signal Process. **118**, 203–210 (2016)
35. Li, C., Lin, D., Lü, J.: Cryptanalyzing an image-scrambling encryption algorithm of pixel bits. IEEE MultiMed. **24**(3), 64–71 (2017)
36. Wen, W., Zhang, Y., Su, M., Zhang, R., Chen, J.-X., Li, M.: Differential attack on a hyper-chaos-based image cryptosystem with a classic bi-modular architecture. Nonlinear Dyn. **87**(1), 383–390 (2017)
37. Fan, H., Li, M., Liu, D., Zhang, E.: Cryptanalysis of a colour image encryption using chaotic APFM nonlinear adaptive filter. Signal Process. **143**, 28–41 (2018)
38. Alanezi, A., Abd-El-Atty, B., Kolivand, H., El-Latif, A., Ahmed, A., El-Rahiem, A., Sankar, S., Khalifa, H.S., et al.: Securing digital images through simple permutation-substitution mechanism in cloud-based smart city environment. Secur. Commun. Netw. **2021** (2021)
39. Li, T., Zhang, D.: Hyperchaotic image encryption based on multiple bit permutation and diffusion. Entropy **23**(5), 510 (2021)
40. Naim, M., Pacha, A.A., Serief, C.: A novel satellite image encryption algorithm based on hyperchaotic systems and josephus problem. Adv. Space Res. **67**(7), 2077–2103 (2021)
41. Yang, Y., Wang, L., Duan, S., Luo, L.: Dynamical analysis and image encryption application of a novel memristive hyperchaotic system. Opt. Laser Technol. **133**, 106553 (2021)
42. Bouridah, M.S., Bouden, T., Yalçin, M.E.: Delayed outputs fractional-order hyperchaotic systems synchronization for images encryption. Multimed. Tools Appl. **80**(10), 14723–14752 (2021)
43. Menezes, A.J., van Oorschot, P.C., Vanstone, S.A.: Handbook of Applied Cryptography (1996)

44. Aumasson, J.-P.: Serious Cryptography: a Practical Introduction to Modern Encryption. No Starch Press (2017)
45. electronics, M.: One time pad encryption the unbreakable encryption method (2016)
46. Li, C., Li, S., Asim, M., Nunez, J., Alvarez, G., Chen, G.: On the security defects of an image encryption scheme. Image Vis. Comput. **27**(9), 1371–1381 (2009)
47. Schmitz, R.: Use of chaotic dynamical systems in cryptography. J. Frank. Inst. **338**(4), 429–441 (2001)
48. Arroyo, D., Diaz, J., Rodriguez, F.: Cryptanalysis of a one round chaos-based substitution permutation network. Signal Process. **93**(5), 1358–1364 (2013)
49. Li, S., Li, C., Chen, G., Bourbakis, N.G., Lo, K.-T.: A general quantitative cryptanalysis of permutation-only multimedia ciphers against plaintext attacks. Signal Process.: Image Commun. **23**(3), 212–223 (2008)
50. Solak, E.: Cryptanalysis of chaotic ciphers. In: Chaos-Based Cryptography, pp. 227–256. Springer (2011)
51. Ott, E.: Chaos in Dynamical Systems. Cambridge University Press (2002)
52. Zhang H., Liu, D., Wang, Z.: Controlling Chaos: suppression, Synchronization and Chaotification. Springer Science & Business Media (2009)
53. Strogatz, S.H.: Nonlinear Dynamics and Chaos with Student Solutions Manual: with Applications to Physics, Biology, Chemistry, and Engineering. CRC Press (2018)
54. Cuomo, K.M., Oppenheim, A.V., Strogatz, S.H.: Synchronization of lorenz-based chaotic circuits with applications to communications. IEEE Trans. Circ. Syst. II: Analog Dig. Signal Process. **40**(10), 626–633 (1993)
55. Pecora, L.M., Carroll, T.L.: Synchronization in chaotic systems. Phys. Rev. Lett. **64**(8), 821 (1990)
56. Vaidyanathan, S., Pakiriswamy, S.: The design of active feedback controllers for the generalized projective synchronization of hyperchaotic Qi and hyperchaotic lorenz systems. In: Computer Information Systems—Analysis and Technologies, pp. 231–238. Springer (2011)
57. Wu, W., Chen, Z.: Hopf bifurcation and intermittent transition to hyperchaos in a novel strong four-dimensional hyperchaotic system. Nonlinear Dyn. **60**(4), 615–630 (2010)
58. Gautschi, W.: Numerical integration of ordinary differential equations based on trigonometric polynomials. Numer. Math. **3**(1), 381–397 (1961)
59. Pano-Azucena, A.D., Tlelo-Cuautle, E., Rodriguez-Gomez, G., de la Fraga, L.G.: FPGA-based implementation of chaotic oscillators by applying the numerical method based on trigonometric polynomials. AIP Adv. **8**(7), 75217 (2018)
60. Chai, X., Fu, X., Gan, Z., Lu, Y., Chen, Y.: A color image cryptosystem based on dynamic DNA encryption and chaos. Signal Process. **155**, 44–62 (2019)
61. Gan, Z.-H., Chai, X.-L., Han, D.-J., Chen, Y.-R.: A chaotic image encryption algorithm based on 3-D bit-plane permutation. Neural Comput. Appl. **31**(11), 7111–7130 (2019)
62. Tsafack, N., Iliyasu, A.M., De Dieu, N.J., Zeric, N.T., Kengne, J., Abd-El-Atty, B., Belazi, A., Abd EL-Latif, A.A.: A memristive rlc oscillator dynamics applied to image encryption. J. Inf. Secur. Appl. **61**, 102944 (2021)
63. Biham, E., Shamir, A.: Differential Cryptanalysis of the Data Encryption Standard. Springer Science & Business Media (2012)

Chaos-Based Image Encryption Based on Bit Level Cubic Shuffling

Lazaros Moysis, Ioannis Kafetzis, Aleksandra Tutueva, Denis Butusov, and Christos Volos

Abstract This work studies the problem of chaos-based image encryption. First, a generalization of the 1D chaotic map proposed by (Talhaoui et al. in The Visual Computer, pp 1–11, 2020) [8] is constructed and studied. The generalized map showcases regions of constant chaotic behaviour, similar to the original map. Based on the new map, a statistically secure pseudo-random bit generator is designed, which is utilised in the encryption process. An image encryption technique based on shuffling the bit levels of an image is introduced, by first arranging the bits in a three dimensional matrix, and performing a three level shuffling, on each individual row, column, and bit level of the 3D matrix. The shuffling is then followed by an exclusive OR operation between the shuffled bits and a bitstream from the proposed chaotic bit generator, which results in the encrypted image. The combination of shuffling and XOR yields a ciphertext image that is resistant to a collection of attacks, like histogram, correlation, and entropy analysis, NPCR and UACI measures, cropping attacks, and is also robust to transmission noise. This is verified by testing the encryption process to a collection of plaintext images. Finally, the encryption/decryption process is implemented in a Graphical User Interface for ease of use.

L. Moysis (✉) · I. Kafetzis · C. Volos
Laboratory of Nonlinear Systems - Circuits & Complexity, Physics Department,
Aristotle University of Thessaloniki, Thessaloniki, Greece
e-mail: lmousis@physics.auth.gr; moysis.lazaros@hotmail.com

I. Kafetzis
e-mail: kafetzis@physics.auth.gr

C. Volos
e-mail: volos@physics.auth.gr

A. Tutueva · D. Butusov
Youth Research Institute, Saint-Petersburg Electrotechnical University 'LETI',
5, Professora Popova st., 197376 Saint Petersburg, Russia
e-mail: avtutueva@etu.ru

D. Butusov
e-mail: dnbutusov@etu.ru

© The Author(s), under exclusive license to Springer Nature Switzerland AG 2022
A. A. Abd El-Latif and C. Volos (eds.), *Cybersecurity*, Studies in Big Data 102,
https://doi.org/10.1007/978-3-030-92166-8_7

157

1 Introduction

Chaos-based cryptography is a well established field in the discipline of information security, with applications spanning secure communications [1], watermarking [2], hashing [3], random bit generators [4], and different types of data masking, like text, sound and image [5, 6].

Chaotic systems combine determinism and unpredictability, with a low computational cost, which makes them an excellent source of randomness, for use in encryption related applications. This is why there is a constant need to develop novel chaotic systems, and especially low dimensional discrete time maps, that can yield large regions of chaotic behaviour [7–10].

Motivated by this, a novel map is constructed, as a generalization of the single-parameter one-dimensional cosine polynomial (1-DCP) chaotic map proposed in [8]. The original map combines a cubic polynomial with the cosine function, to yield a map with complex chaotic behaviour. The map showcased large regions of chaotic behavior, which makes it suitable for use in encryption designs. On the other hand, the map only had a single parameter, so its key space was not sufficient, to resist brute force attacks. These two properties motivated the consideration of a modified version of this map.

Hence, the map proposed in the current work is a generalization of the (1-DCP) map, with a second parameter added inside the cosine function. The new map has a larger key space, since it consists of two parameters. The analysis of the map is performed through computation of its bifurcation diagrams and Lyapunov exponent, and it is seen that the map has regions of constant chaotic behaviour, which is desired for encryption related applications. Also, the phase diagram for large parameter values has no distinguishable shape, which is an additional desired property for cryptographic applications. Moreover, the computational load of the new map is practically the same to the original, since they differ by a single operation of addition.

Based on the proposed map, a statistically secure pseudo-random bit generator (PRBG) is developed. PRBGs are a very common application for chaotic systems, since they are used as a basis in most encryption schemes. For example, in a recent work [4] a PRBG is proposed based on a delayed Chebysev map, that can generate 8 bits per iteration. In [11], another PRBG that generates 8 bits per iteration is proposed, based on a modified logistic map. In [12], the technique of bit reversal is applied to increase complexity in the PRBG. In [13], adaptive chaotic maps are proposed that can be used to effectively increase the key space of a PRBG. The literature on PRBGs is extensive, so for an in depth presentation, the recent survey [14] considers a large set of research works, and compares their performance. The PRBG constructed here applies commonly used techniques to generate the bits, like the modulo 2 operator. One difference from other methods though is the use of a delay term in the bit generation, which can increase the generator's randomness.

After the PRBG is designed, the problem of image encryption is considered. The application of chaotic systems in image encryption is a well established and expanding research area, with many developed techniques. For a guide, the survey [6] reviews different approaches to chaos-based image encryption. As recent examples of diverse approaches to image encryption, in [15] a technique was proposed based on Barnsleyćs chaos game, and a Graphical User Interface was also developed. In [9] a two step image encryption was proposed based on a novel chaotic map, where the pixels are first shuffled, and then modulated with the values of the chaotic map. In [16], the technique of DNA encoding was applied to medical images. In [17], a continuous chaotic system was used to design an S-box for image encryption. In [18], a continuous hyperchaotic system was applied for image encryption, in combination with a zigzag operation to scramble the image pixels.

From the above works, it can be concluded that the best chaos encryption schemes need to combine the operations of confusion and diffusion. Confusion is the process of hiding the connection between the bits of the original (plaintext) image and the encrypted (ciphertext) image. Confusion can be achieved is through substitution of the image's binary information. The most common operator to achieve this is the exclusive OR operator (XOR) and this is where chaos based PRBGs are utilized, as they can be combined with the plaintext image bits to mask them. Diffusion is the operation of making the encrypted image sensitive to changes in the plaintext image. A way to achieve diffusion is to make the encryption keys plaintext dependent. Moreover the proccess of rearranging the pixels of the plaintext image is beneficial in reducing pixel correlation, and also spreading out the key pixel information through-out the image. This can be achieved by generating shuffling rules chaotically. An even more advanced approach is to perform this operation on the binary level, across all binary levels of the image's pixels. This process not only reduces correlation, but also rearranges the most significant bits of each pixel across all bit levels and pixel positions, which can make the encryption much more secure, and also resistant to loss of information or corruption during the transmission stage.

Motivated by the above key aspects of encryption, a method is developed for chaotic encryption, giving an emphasis to the process of thoroughly permuting the binary information of the plaintext image. In the technique proposed in this work, a rearranging of the bits across all rows, columns, and bit levels of the matrix is performed. Such an approach can yield increased security, as well as resistance to noise and cropping attacks, since the bits that carry most of the image information are redistributed across all bit levels. This is why many recent techniques consider permutation and encryption on the bit levels of a plaintext image. As recent examples, in [19–22] the image pixels are shuffled to reduce correlation in the image. In [23] a technique for colour images was proposed, through a combination of a continuous and a discrete chaotic system, to perform a combined shifting of the pixels and bits of its three RGB channels. In [24], a modified pulsed-coupled spiking neurons circuit map is used to encrypt an image, using a novel technique of breaking each pixel into four bit pairs, and rearranging them to form a new matrix of double size compared to the original image. In [25], colour image encryption is performed by simultaneously shuffling the bit levels of all three colour sub-images, while again both discrete and continuous systems are used for the complete encryption process. In [26], a self-adaptive bit level encryption is proposed, where an image is broken down into parts, and one part is used to encrypt the other. In [27], quadratic and cubic maps are used to shuffle the bits and encrypt a chosen number of bit levels. In [28], bit level permutation is performed based on the popular Rubik's cube game. In [29], shuffling is performed using the Hilbert curve pattern and cyclic shift.

The procedure developed here consists of two main steps. First, the pixels of a given plaintext image are decomposed into 8 bits and arranged in a 3D matrix. Then, each individual row level, column level, and pixel level of the 3D matrix is shuffled. The shuffling rules for each level are generated from the values of three different chaotic maps. The rearrangement of the bits results in the dispersion of the significant bits of the plaintext across all the eight pixel levels and all the rows and columns of the image. This yields a shuffled image with no distinguishable information, as the value of each pixel after the shuffling is changed. Moreover, in addition to showing security against attacks, like correlation and entropy analysis, such a thorough bit permutation makes the image resistant to loss of information and noise, as will be seen in the simulations section. The trade-off for achieving this is an increased cost in computational time.

Moreover, since only rearranging the bits of the plaintext cannot yield security against specific attacks, like known plaintext attacks, a second step of confusion is added in the design. Here, the bits of the shuffled image are combined with a chaotic bitstream of the same length, generated from the proposed PRBG, using the XOR operation. This results in a ciphertext image that is secure against known plaintext attacks as well.

Also, to further improve the resistance to known plaintext attacks, the encryption keys used are plaintext-dependent. Two general approaches for this is to either use a custom rule to generate the keys based on the pixel values of the image [30–35], or to use hash functions [16, 36–40]. Here, the first approach is taken, so the parameter values and initial conditions of the maps used for the shuffling and PRBG are computed based on the plaintext image to be encrypted.

To verify the security of the design, the encryption is performed to a collection of plaintext images, which are then submitted to a series of statistical tests and measures. These include histogram analysis, correlation analysis, global and local information entropy measure, differential attack analysis, cropping and noise attack analysis, sensitivity analysis, as well as resistance to brute force attacks. All of the performed tests verified that the design is indeed secure to all the types of attacks considered.

Finally, the proposed scheme is implemented with a Graphical User Interface (GUI), which is a valuable tool for easily implementing and visualising the procedure. This tool can be picked up by any interested reader that can use the proposed design to encrypt personal data.

Overall, the key aspects of the proposed work can be outlined as follows:

1. A novel chaotic map with increased key space and wide regions of chaotic behavior is proposed.
2. A simple, statistically secure PRBG is designed based on the chaotic map.
3. An image encryption technique is developed, with confusion and diffusion steps. The technique gives emphasis on the permutation of each pixel bits accross all

rows and columns of the image. The resulting method was tested against a collection of statistical tests and attacks, yielding resistance to all of them.
4. The overall design was implemented in a GUI, which facilitates its usage.

The rest of the chapter is structured as follows: Sect. 2 presents the generalized version of the map in [8] and studies its dynamical behaviour. In Sect. 3, a PRBG is constructed using the proposed map, and its statistical randomness is verified. Section 4 presents the complete encryption design. In Sect. 5, a collection of plaintext images are encrypted, and then submitted to a series of tests and measures, to evaluate the design. In Sect. 6, the complete procedure is implemented in a Graphical User Interface. Finally, Sect. 7 concludes the work, with a discussion on future topics of interest.

2 The Proposed Chaotic Map

Recently in [8], Talhaoui, Wang & Midoun proposed the following one-dimensional cosine polynomial chaotic map

$$x_k = \cos(\mu(x_{k-1}^3 + x_{k-1})) \tag{1}$$

where $\mu > 0$ is a parameter. The map exhibits chaotic behaviour for almost all parameter values, especially higher ones. This can be verified by the two bifurcation diagrams of (1) with respect to parameter μ, which are shown in Fig. 1, for low and high values of the parameter, and initial condition $x_0 = 0.1$. For lower values, the map exhibits the common phenomenon of period doubling route to chaos, as well as crisis phenomena, where it abruptly exits chaos, as can be seen around $\mu = 1.5$. For higher parameter values however, wide ranges of chaotic behaviour are observed. Also, since the map utilises a cosine function, the values are mapped on the interval $[-1, 1]$.

Based on this map, a simple generalization is given by:

$$x_k = \cos(\mu(x_{k-1}^3 + x_{k-1}) + a) \tag{2}$$

where $a > 0$ is a second parameter that shifts the argument of the cosine function. Due to the periodicity of the cosine function, this parameter is considered in the interval $a \in [0, 2\pi)$. Clearly, the original map (1) is a special case of (2) for $a = 0$. Two bifurcation diagrams of (2) are shown in Fig. 2 with respect to parameter μ, under the same low and high value ranges, when the value of the second parameter is $a = 4$, and $x_0 = 0.1$. Compared to the original map, it can be seen that the behaviour of the new map showcases similar chaotic phenomena for low values of μ, and remains consistenly chaotic for higher values.

The above result can be verified by also considering the bifurcation diagram with respect to a for the proposed map. Figure 3 shows the diagram for a in the range

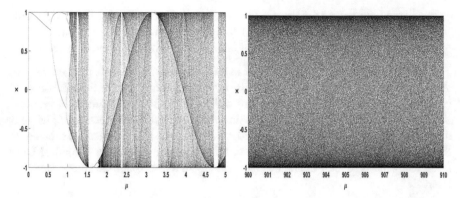

Fig. 1 Bifurcation diagram of the map (1) for two different ranges of the parameter μ and $x_0 = 0.1$

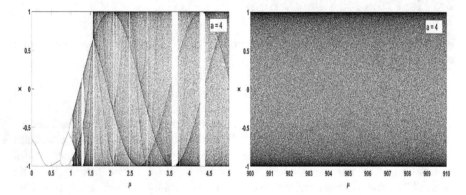

Fig. 2 Bifurcation diagram of the proposed map (2) for two different ranges of the parameter μ and $a = 4$, $x_0 = 0.1$

$[0, 2\pi)$, for two different values of the parameter μ. For $\mu = 1.5$, the map showcases regions of constant chaos, as well as abrupt changes in the shape of the attractor, as can be seen in the range $a \in (3, 2\pi)$. For a higher value of μ though, the system retains its constant chaotic behaviour for all values of a.

For a clearer visualisation of the behaviour of (2) for small values of μ, Fig. 4 shows a graph indicating the chaotic and non-chaotic pairs of parameters (μ, a). Here, interesting patterns emerge, showcasing the complex interchange between chaotic and periodic behaviour.

Also, Fig. 5 shows a comparison between the Lyapunov exponent (LE) of the original map (1) and the proposed map (2) for $a = 4$. Here, it can be seen that the two maps maintain practically the same value of LE in the computed range. Overall, the proposed map can maintain the chaotic behaviour of the original map, and since it has an additional parameter, it has an increased key space, which can make it useful in encryption related applications [41].

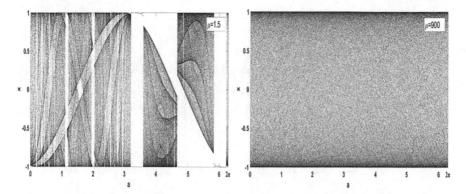

Fig. 3 Bifurcation diagram of the proposed map (2) for two different ranges of the parameter a and $\mu = 1.5$, $\mu = 900$, $x_0 = 0.1$

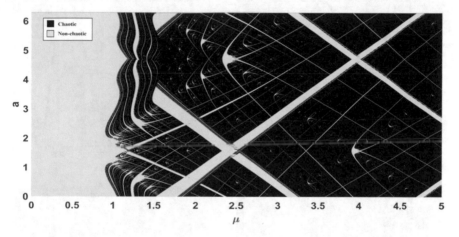

Fig. 4 Pairs of parameter values (μ, a) of the map (2) that yield chaotic (black) and non-chaotic (yellow) behaviour

Finally, Fig. 6 shows different phase diagrams for the original and modified map, under different parameter values. For both maps, it can be seen that for higher values of the parameter μ, the phase diagram does not have a distinguishable shape, a characteristic that is desired for maps used in encryption related applications [4, 12].

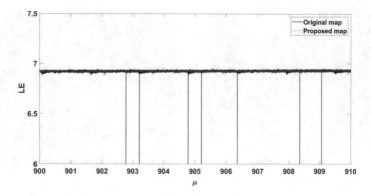

Fig. 5 Diagram of Lyapunov exponent of (1) and (2) for $a = 4$, $x_0 = 0.1$

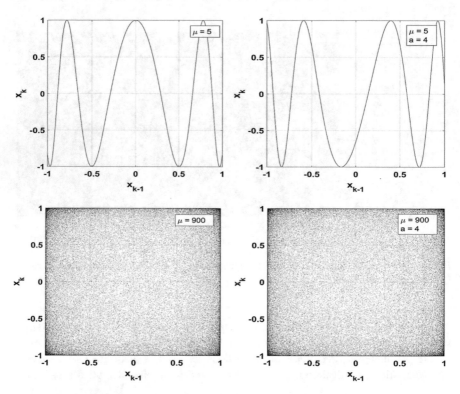

Fig. 6 Phase diagrams of the original (1) (left) and modified (2) (right) maps, for different parameter values, and $x_0 = 0.1$

3 Construction of Pseudo-Random Bit Generator

In this section, a PRBG is designed based on the map (2). Using the values of the map, the bits are generated based on the following rule

$$b_k = \lfloor mod(10^{12}|x_k + x_{k-1}|, 2) \rfloor \qquad (3)$$

where $\lfloor \cdot \rfloor$ denotes the floor operation. The resulting bitstream is $\mathcal{B} = \{b_0, b_1, ...\}$. The use of the sum $|x_k + x_{k-1}|$ helps in increasing the complexity of the PRBG. To verify the statistical randomness of the generator, a set of $100 \cdot 10^6$ bitstreams is generated and tested through the National Institute of Standards and Technology (NIST) test suite [42]. The package consists of 15 statistical tests. Each test is used to verify the randomness of a bitstream. For each test, a P-value is returned, and if it is higher than a significance level, chosen here as the default 0.01, the test is considered succesful. For a generator to be considered random, all the tests should be successful. The results are shown in Table 1, where it can be seen that the generator passes all the tests, and can thus be used in an encryption application. Note that for tests that have multiple sub-cases, like the NonOverlappingTemplate, only the P-value of the last case is printed.

Table 1 NIST test results ($x_0 = 0.1$, $\mu = 900$, $a = 4$)

No.	Test	Chi-square P-value	Rate
1	Frequency	0.779188	99/100
2	BlockFrequency	0.779188	100/100
3	CumulativeSums	0.275709	99/100
4	Runs	0.779188	99/100
5	LongestRun	0.678686	98/100
6	Rank	0.096578	98/100
7	FFT	0.851383	98/100
8	NonOverlappingTemplate	0.017912	100/100
9	OverlappingTemplate	0.122325	100/100
10	Universal	0.419021	99/100
11	ApproximateEntropy	0.514124	100/100
12	RandomExcursions	0.021999	56/57
13	RandomExcursionsVariant	0.759756	57/57
14	Serial	0.419021	99/100
15	LinearComplexity	0.911413	98/100

4   The Proposed Shuffling and Encryption Technique

4.1   Image Bit Levels

In a grayscale image A, every pixel has a grey intensity between black and white, which is represented by an integer between 0 and 255, with 0 corresponding to the black value, 255 to the white, and all integers in between representing the shades of grey. Since each integer in the interval [0, 255] can be represented using 8 bits, every pixel value a_{ij} can be represented in binary format as:

$$a_{ij} = a_{ij}^1 2^0 + a_{ij}^2 2^1 + a_{ij}^3 2^2 + a_{ij}^4 2^3 + a_{ij}^5 2^4 + a_{ij}^6 2^5 + a_{ij}^7 2^6 + a_{ij}^8 2^7 \qquad (4)$$

where $a_{ij}^w \in \{0, 1\}$ denotes the value of the w-th bit. By representing each pixel in the above format, an image can be broken down into eight individual binary subimages, where each image has the value a_{ij}^w at the (i, j) position. These binary images are called pixel levels, or pixel planes of an image, as shown in Fig. 7. An example of the different pixel levels of a grayscale image is shown in Fig. 8. By observing the different bit levels, it is clear that lower levels do not contain any distinguishable information, while higher levels carry more information about the image. This is undestandable if one considers (4). A bit value of $a_{ij}^1 = 1$ at the first level will only contribute the value of $2^0 = 1$ on the pixel value a_{ij}, while a bit value of $a_{ij}^8 = 1$ at the eighth level will contribute the value $2^7 = 128$ on the pixel value a_{ij}. Thus, bits at the higher levels have a greater effect on the final pixel value, and thus carry more information about the image [24–26]. The percentage of pixel information for each bit level is given by:

Fig. 7 Bit levels of a grayscale image

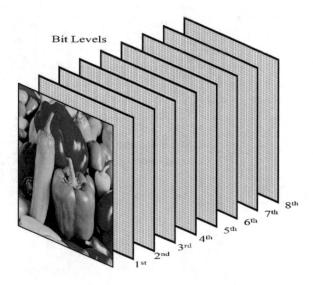

Bit Levels

1st 2nd 3rd 4th 5th 6th 7th 8th

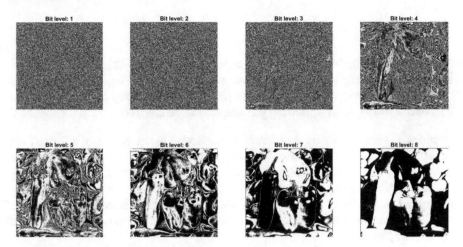

Fig. 8 The individual bit levels of a grayscale image

Table 2 Percentage of pixel information carried in each bit level

Level	Percentage (%)
$p(1)$	0.3921
$p(2)$	0.7843
$p(3)$	1.5686
$p(4)$	3.1373
$p(5)$	6.2745
$p(6)$	12.5490
$p(7)$	25.0980
$p(8)$	50.1961

$$p(i) = \frac{2^{i-1}}{\sum_{j=1}^{8} 2^{j-1}} 100\%, \quad i = 1, ..., 8 \tag{5}$$

and the results are shown in Table 2. It can be verified that the last four levels carry over 90% of the image information. As will be seen in the following section, a shuffling of the bits among bit levels allows the significant information of a pixel to be distributed along all pixel levels and positions. This not only makes the design more secure, but also allows for reconstruction of the image in case information is lost or corrupted during transmission.

4.2   Encryption Algorithm

The encryption algorithm utilises the proposed chaotic map (2) and the PRBG of Sect. 3, and consists of two main steps, a three level shuffling operation and an exclusive OR operation. Algorithm 1 is a thorough, step-by-step presentation of the proposed method. In the following paragraphs, the outline of each step is explained.

Remark 1 First, a note on notation. When referring to the entries of a matrix, the ":" symbol denotes all the entries with respect to the given dimension. For example for a $M \times N$ matrix A, the term $A(1, :)$ refers to the entries of the first row. Similarly, for a $M \times N \times 8$ matrix B, the term $B(1, :, :)$ denotes the $N \times 8$ matrix formed by taking the entries in all the N columns and all 8 levels that correspond to the first row. Note that this syntax is adopted from MATLAB. Such sub-matrices are also depicted in Fig. 9.

Also, in Algorithm 1, the subscript i denotes rows, j denotes columns, w denotes bit levels, and k denotes iterations of the chaotic map used in each Step.

There are four chaotic maps of the form (2) used, three in the shuffling process, and one to generate the PRBG. The first step is to compute their key parameters. The parameters are dependent on the plaintext image used, and are computed using (15). Having the key values being plaintext dependent can make the encryption scheme resistant to plaintext attacks [16, 30–34, 36, 37].

After the key values are chosen, the $M \times N$ input image $\mathcal{A}$ is decomposed into its bit levels, resulting in a 3D binary matrix representation $\mathcal{A}_{bin}$ of dimensions $M \times N \times 8$. So an element a_{ij}^w of matrix $\mathcal{A}_{bin}$ represents the w-th bit of the pixel in the position (i, j).

Once the 3D binary matrix $\mathcal{A}_{bin}$ is obtained, the three level shuffling procedure begins. In essence, this procedure consists of performing a shuffling operation individually in each row level, column level, and pixel level of the matrix. The three levels are depicted in Fig. 9. Similar shuffling techniques appear in [19–21], but they are applied to the pixels of the original $M \times N$ image, not its binary representation.

Starting from the row levels, the matrix $\mathcal{A}_{bin}$ is decomposed into M different $N \times 8$ matrices, which correspond to taking all the columns and bit levels for an individual matrix row, that is $\mathcal{A}_{bin}(1, :, :)$, $\mathcal{A}_{bin}(2, :, :)$,..., $\mathcal{A}_{bin}(M, :, :)$. The rows and columns of each individual matrix are then shuffled, via left and right matrix multiplication by invertible matrices, generated using the first chaotic map. This process is analytically described in Steps 3a.-3c. of Algorithm 1. After the shuffling of all individual row levels is complete, the resulting matrix is denoted as $\mathcal{R}$.

Then, the second shuffling is performed, this time on the column levels of $\mathcal{R}$. So the 3D matrix $\mathcal{R}$ is decomposed into N different $M \times 8$ matrices, which correspond to taking all the rows and bit levels for an individual column, that is $\mathcal{R}(:, 1, :)$, $\mathcal{R}(:, 2, :)$,..., $\mathcal{R}(:, N, :)$. Each individual matrix is then shuffled using the same technique described in the previous step, and in Steps 4a.–4c. of Algorithm 1, using the second chaotic map. This process results in matrix C.

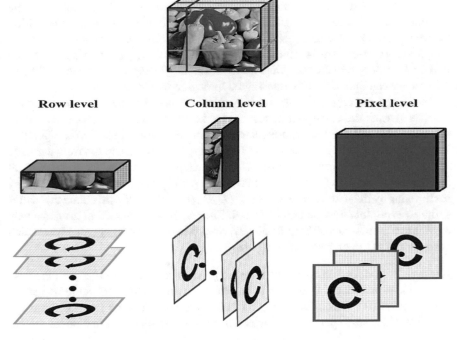

Fig. 9 The three levels of shuffling performed on the $M \times N \times 8$ 3D matrix that represents the image

Similarly, the third shuffling is performed by decomposing matrix C into its eight $M \times N$ pixel levels $C(:, :, 1)$, $C(:, :, 2)$,..., $C(:, :, 8)$. Each individual level is then shuffled using the above technique, as described in Steps 5a.–5c. of Algorithm 1, using the third chaotic map. The resulting 3D matrix is denoted as $\mathcal{P}$.

The three levels of shuffling result in the 3D binary matrix $\mathcal{P}$, which, using (4), can be reversed into a grayscale image. Since the shuffling procedure is performed on all bit levels, the resulting image has different pixel values than the original plaintext image $\mathcal{A}$, so it is actually a ciphertext of the original image. Yet, as will be seen in the next Section, performing only the shuffling procedure will not guarantee security against all types of attacks. For example, an all black image where each pixel value is zero will result in a 3D binary matrix consisting only of zeros, thus the shuffling of the bits will have no effect on the plaintext. This will make the design vulnerable to known plaintext attacks. Hence, the shuffling procedure is followed by a second encryption step, where an exclusive OR operation is performed.

For this second round of encryption, the bits of the $M \times N \times 8$ matrix $\mathcal{P}$ are reshaped into a vector of length $M \cdot N \cdot 8$, and are combined using XOR with a bitstream of the same length, generated using the fourth chaotic map and the PRBG of the previous section. The resulting bitstream denoted as $\mathcal{E}_{stream}$ can then be reshaped into an $M \times N \times 8$ matrix $\mathcal{E}_{bin}$, and transformed into the ciphertext image using (4). This image can then be safely transmitted through a public channel.

To decrypt the cihpertext, the reverse procedure needs to be followed. Once the receiver obtains the ciphertext image and the key values for the maps, each step of Algorithm 1 is followed backwards, starting from the last one. Initially, the ciphertext image is reshaped into a vector and $X O R$ed using the same bitstream from the same PRBG used for the encryption process. Then, Step 5 is reversed by computing all the shuffling matrices for each pixel level and performing the reverse shuffling by multiplying by the matrix inverses as $L_w^{-1}(L_w \mathcal{P}_w Q_w) Q_w^{-1} = \mathcal{P}_w$, to obtain the matrix C that was produced at the end of Step 4. Then Step 4 is reversed in the same way to obtain $\mathcal{R}$ at the end of Step 3. Finally, Step 3 is reversed to obtain $\mathcal{A}_{bin}$, which corresponds to the original plaintext image.

Remark 2 In determining the plaintext dependent key values in (15), in the case where the sum inside the modulo is an integer, then $\lambda = 0$. This could happen in known plaintext attacks, for example in all white and all black images. Choosing $\lambda = 0$ would lead to periodic behaviour in the chaotic maps, so to counter this, in case $\lambda = 0$, the parameter is replaced by $\lambda = mod\left(\log_e \left(M \cdot N + \frac{\sum_{i=1}^{M} \sum_{j=1}^{N} a_{ij}}{M \cdot N} \right) + Ent(A), 1 \right)$.

5 Encryption Performance

The proposed technique is applied to seven different grayscale images taken from the USC-SIPI Image Database (http://sipi.usc.edu/database/), namely the peppers, baboon, airplane, earth, gray shades, all black, and all white images. The encryption results are shown in Fig. 10 where the original (plaintext), shuffled, and encrypted (ciphertext) images are shown. In all cases, the encrypted image visually displays no recognisable information on the original plaintext. This is also the case for the shuffled images, except from the case of the all black and all white images, where the shuffling has no effect. In the following, the performance of the encryption scheme is tested with respect to different measures, to verify that the encrypted images are indeed secure against attacks that try to unmask information on the plaintext image.

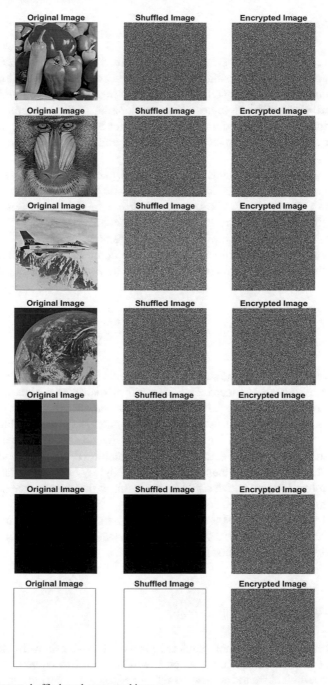

Fig. 10 Plaintext shuffled, and encrypted images

5.1 Histogram

The simplest statistical analysis that can be performed on an image is to measure the distribution of all the 256 possible grey values. The greyscale distribution is plotted in a histogram. An image depicting any sort of information will naturally have an uneven distribution of pixel values. Thus, by looking at the histogram, it is easy to identify images carrying information, despite not being identifiable to the naked eye. So, for a ciphertext to be secure to statistical attacks, it should have a uniform histogram.

The histograms of the plaintext, shuffled and encrypted images are shown in Fig. 11. Looking at the shuffled images first, it can be observed that the histogram is not uniform in all cases, as can be seen in the Airplane and Shades cases. As already pointed out, the shuffling has no effect on the monocoloured images. In the encrypted case though, the histogram is uniform for all image files.

In addition, the variance of each histogram is computed, to obtain a statistical measure of the histogram's uniformity, apart from its visual inspection [43, 44]. The variance is computed as

$$var(\mathcal{H}) = \frac{1}{256^2} \sum_{i=0}^{255} \sum_{i=0}^{255} \frac{(h_i - h_j)^2}{2} \tag{6}$$

where $\mathcal{H} = \{h_0, ..., h_{255}\}$ is the sum of measurements for each grayscale value in the image pixels. A smaller variance indicates a uniform distribution of the pixel values. The results are shown In Table 3. As can be seen, the variance of the encrypted image histograms is lower than the plaintext and shuffled ones. In the same Table, results of recent works are also printed, for the images that are considered in each study. The variance results are comparable to these works, meaning that the method is equally efficient. Notice though that other works reported slightly different variances for the original plaintext images, which could be the result of different source databases or different file types used.

Overall, by studying the histogram of the encrypted images, no information about the pixel value distribution for the plaintext image can be extracted, and the scheme can be considered secure against histogram analysis.

5.2 Correlation

In images depicting information, adjacent pixels will have relatively close values. This means that their correlation coefficient will be high. In a ciphertext image, the correlation coefficient should be close to zero for adjacent pixels, indicating that their values are uncorrelated. The correlation coefficient γ is computed by:

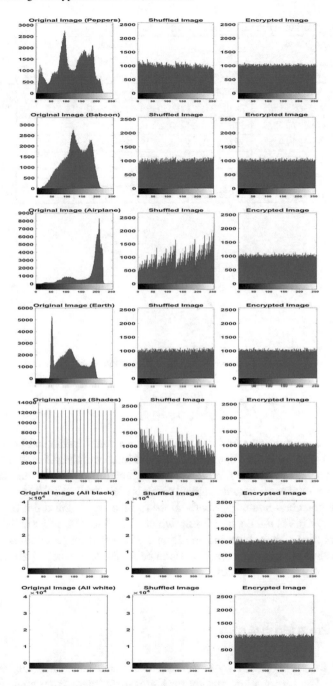

Fig. 11 Histograms of the plaintext, shuffled, and encrypted images

Table 3 Histogram variance for plaintext, shuffled and encrypted images

	Proposed			Safack et al. [43]	Abdelfatah [44]	Nepomuceno et al. [35]
	Original	Shuffled	Encrypted	Encrypted	Encrypted	Encrypted
Peppers	482548.1411	4697.4588	894.7686	–	1076.6	–
Baboon	752365.2156	2846.5411	1095.1058	909.4980	880.2	990.11
Airplane	2882376.6274	97676.7215	1275.6156	–	–	–
Earth	1066350.9254	1670.9098	956.5019	–	–	–
Shades	11780472.3294	63914.9254	1242.5568	–	–	–
All black	268435456	268435456	1091.9215	–	–	–
All white	268435456	268435456	996.2823	–	–	–

$$E(x) = \frac{1}{N} \sum_{i=1}^{N} x_i, \tag{7}$$

$$D(x) = \frac{1}{N} \sum_{i=1}^{N} (x_i - E(x))^2, \tag{8}$$

$$cov(x, y) = \frac{1}{N} \sum_{i=1}^{N} (x_i - E(x))(y_i - E(y)), \tag{9}$$

$$\gamma(x, y) = \frac{cov(x, y)}{\sqrt{D(x)}\sqrt{D(y)}}, \tag{10}$$

where x, y are the gray values of two adjacent pixels, and N is the number of adjacent pairs of pixels (x, y). Table 4 displays the correlation coefficients for all the plaintext, shuffled and encrypted images. While for the original images the correlation between adjacent horizontal, vertical and diagonal pixels is high, for the shuffled and encrypted images, the correlation is consistently close to zero, up to the order of magnitude $10^{-3} - 10^{-4}$. This is the desired result, which is in line with most recent works in image encryption. Thus, both the shuffling and encryption procedures manage to reduce correlation, with the exception again being the all black and all white cases, for which shuffling has no effect.

5.3 Information Entropy

The global information entropy is a measure of randomness and unpredictability for a signal. The information entropy for a grayscale image is computed using the technique given in [5]:

Table 4 Correlation coefficients for the image encryption

	Original			Shuffled			Encrypted		
	Horizontal	Vertical	Diagonal	Horizontal	Vertical	Diagonal	Horizontal	Vertical	Diagonal
Peppers	0.9768	0.9792	0.9639	0.0003	0.0022	0.0021	0.0005	0.0002	0.0007
Baboon	0.8665	0.7587	0.7262	0.002	−0.0022	−0.0005	−0.0006	−0.0006	0.0001
Airplane	0.9663	0.9641	0.9370	0.0030	0.0079	0.0032	−0.0002	−0.0007	−0.0020
Earth	0.9709	0.9761	0.9515	0.0017	0.0022	−0.0014	−0.0003	−0.0049	−0.0014
Shades	0.9965	0.9998	0.9964	0.0022	0.0063	0.0003	−0.0020	−0.0002	0.0002
All black	–	–	–	–	–	–	0.0009	0.0037	−0.0024
All white	–	–	–	–	–	–	−0.0018	−0.0010	0.0008

$$H(S) = -\sum_{i=0}^{2^8-1} p(s_i) \log_2 p(s_i), \tag{11}$$

where $p(s_i)$ is the probability of occurrence for the value s_i. The information entropy of an encrypted image should be close to 8, which indicates randomness in the expected pixel values. Table 5 shows the entropy of the original plaintext, shuffled and encrypted images. In all cases, the encrypted images have the highest entropy value, close to the ideal value of 8, which indicates that the resulting ciphertext is secure to entropy checks. This is aligned with recent works on image encryption.

An additional entropy measure that can be used to test the randomness of an image is that of local entropy. In contrast to global entropy described above, which measures entropy over the complete image, local entropy considers individual sub-regions of the image. It is computed by considering a set of randomly chosen non-overlapping blocks of an image, and computing the average entropy across all blocks [45–50]. Here a set of 30 randomly chosen non-overlapping blocks of size 44×44 is considered, so each block consists of $44^2 = 1936$ pixels. The results are shown in Table 6. In all cases, the encrypted images maintain high values of local entropy around 7.9, in contrast to the plaintext images. The results are all comparable to recent works which also considered the local entropy.

Table 5 Entropy for plaintext and encrypted images

	Original	Shuffled	Encrypted
Peppers	7.5937	7.9968	7.9994
Baboon	7.3583	7.9980	7.9993
Airplane	6.7025	7.9355	7.9991
Earth	7.1530	7.9989	7.9993
Shades	4.3923	7.9575	7.9991
All black	0	0	7.9993
All white	0	0	7.9993

Table 6 Local entropy for plaintext and encrypted images

	Proposed			Chen et al. [48]	Sambas et al. [49]	Safack et al. [43]	Alanezi et al. [51]
	Original	Shuffled	Encrypted	Encrypted	Encrypted	Encrypted	Encrypted
Peppers	6.2355	7.9077	7.9060	7.9030	7.9027	7.9034	7.9014
Baboon	6.6679	7.9009	7.9092	7.9022	7.9020	7.9030	7.9032
Airplane	5.3389	7.8392	7.9048	–	7.9033	7.9009	7.9022
Earth	6.2019	7.9088	7.9078	–	–	–	–
Shades	0.3694	7.8624	7.9075	–	–	–	–
All black	0	0	7.9044	–	–	–	–
All white	0	0	7.9070	–	–	–	–

5.4 Differential Attack Analysis

In an encryption scheme, a small alteration of the original plaintext image should lead to a completely different encrypted image. This makes the design secure to known plaintext attacks, where an attacker chooses to encrypt the same plaintext repeatedly, changing it slightly each time, and observe changes in the ciphertext, in order to unmask the structure of the encryption.

There are two measures to identify the degree of change between almost identical plaintexts, called the number of pixels change rate (NPCR) and the unified average changing intensity (UACI). To compute these measures, two identical images are considered, where the second one differs from the first in only a single pixel. Both images are encrypted, and the above measures are computed by the following equations [5]:

$$\text{NPCR} = \frac{\sum_{i=1}^{M} \sum_{j=1}^{N} D_{ij}}{MN} 100\% \tag{12}$$

$$\text{UACI} = \frac{\sum_{i=1}^{M} \sum_{j=1}^{N} |\mathcal{E}_{ij} - \hat{\mathcal{E}}_{ij}|}{MN255} 100\% \tag{13}$$

where $\mathcal{E}$ and $\hat{\mathcal{E}}$ are the two encrypted images, and

$$D_{ij} = \begin{cases} 0 & \mathcal{E}_{ij} = \hat{\mathcal{E}}_{ij} \\ 1 & \mathcal{E}_{ij} \neq \hat{\mathcal{E}}_{ij} \end{cases} \tag{14}$$

The ideal values are 99.61% for the NPCR and 33.46% for the UACI, so the closer the real values are to these, the stronger the encryption design. Since in the proposed methodology, the key values for the maps used for shuffling and XOR are dependent

Table 7 NPCR and UACI measures for different images

	Shuffled		Encrypted	
	NPCR (%)	UACI (%)	NPCR (%)	UACI(%)
Peppers	99.6063	33.4544	99.6239	33.4492
Baboon	99.5949	33.4303	99.6105	33.4467
Airplane	99.5823	33.0876	99.6056	33.4818
Earth	99.6037	33.5053	99.6216	33.4195
Shades	99.5796	33.3158	99.6048	33.4314
All black	0.0004	0.0002	99.6170	33.4088
All white	0.0004	0.00004	99.6048	33.4611

on the plaintext image, changing it even by one pixel will change the key values and thus yield a different shuffled and ciphertext image. Table 7 shows the NPCR and UACI values for all images. It can be seen that both the shuffled and encrypted images perform satisfactorily with respect to both measures, as the results are close to the ideal values, which is also the case for recent works in image encryption. Note that the shuffled all black and all white images yield an NPCR and UACI measure close to, but not identical to zero, as in the case of the edited image with one changed pixel, the keys are computed from (15) and not as in Remark 2, so the shuffling rules are changed.

5.5 Cropping and Noise Attacks

During transmission, it is possible that some part of the transmitted message may be lost, or corrupted. Hence, it is essential to evaluate to what extend it is possible to reconstruct an image when the ciphertext is either cropped, tempered with, or affected by noise [10, 15, 22, 33, 34, 52–54]. Figure 12 shows the result of decryption that is performed from a ciphertext that was cropped, and Fig. 13 the result when the ciphertext is tampered with. In all cases, it can be seen that despite the loss of information, the encrypted image can be adequately reconstructed, even in cases where 50% of the ciphertext has been cropped or tampered with. When the distortion reaches 75%, the outlines of the objects are still visible.

The resistance to cropping attacks is attributed to the fact that during the shuffling process, the significant bits on the 5th–8th levels are redistributed across different pixel levels and positions in the resulting shuffled image. Thus, cropping out a specific portion of the ciphertext does not correspond to loss of information for the identical portion of the plaintext.

Fig. 12 Decryption from different levels of cropped ciphertext images

Fig. 13 Decryption from different levels of tampered ciphertext images

Similarly, Fig. 14 shows the decryption performed from ciphertext corrupted by salt and pepper noise of different intensities. Again, the reconstructed image is clear enough to identify the original ciphertext, even when 50% of the pixels are corrupted. At 75% of corruption, most of the information is lost, but still the outlines of the objects are visible, so some information is saved.

Fig. 14 Decryption from different levels of noisy ciphertext images

Table 8 Comparison of key space

Work	Key space
Proposed	2^{640}
Irani et al. [22]	2^{186}
Chen et al. [48]	2^{193}
Sambas et al. [45]	2^{213}
Liu and Miao [9]	2^{277}
Nepomuceno et al. [35]	2^{283}
Safack et al. [43]	2^{478}
Belazi et al. [16]	2^{716}
Abdelfatah [44]	2^{772}

5.6 Key Space

Any encryption system should be resistant against brute force attacks. For this, it is required that the key space be higher than 2^{100} [41]. In the proposed system, four chaotic maps (2) are used, with parameters $x_0, \mu_1, a_1, ,y_0, \mu_2, a_2, z_0, \mu_3, a_3, w_0, \mu_4, a_4$. Hence, there are overall 12 parameters. Assuming a 16 digit accuracy, an upper bound for the key space can be computed as $10^{12.16} = 10^{192} = (10^3)^{64} \approx (2^{10})^{64} = 2^{640}$. This is higher than the required bound. Table 8 shows a comparison between the key spaces of recent works. It can be seen that the proposed work has among the highest key spaces.

Moreover, for the case of all black and all white images where the shuffling has no effect, the key space reduces to the key space of the XOR operation. Since 3 parameters are required, the key space upper bound is $10^{3.16} \approx 2^{160}$, which again is higher than the required threshold.

Note that since all 12 key parameters are computed based on the plaintext dependent value λ, if an adversary had complete knowledge on the design and the way the key parameters are computed in (15), they would be able to break the encryption by performing a brute force attack just on parameter λ. This would make the design vulnerable, but this issue can be bypassed by small modifications of the procedure. For example, the plaintext can be broken down into four subfigures, and each subfigure can be used to compute an independent key $\lambda_1, \lambda_2, \lambda_3, \lambda_4$, similarly to λ. Then the λ_i can be used to compute the keys for each individual chaotic map as in (15).

5.7 Key Sensitivity

Any encryption scheme should be sensitive to changes in the encryption key, meaning that a small change in any of the key values would lead to a failure of the decryption process [45, 47, 51, 55]. This is the case in our proposed design as well. Since chaotic systems are used for encryption, the process inherits the key sensitivity of the chaotic maps that are used in its source. So any change in the parameter values leads to deviating chaotic trajectories, which makes the encrypted image impossible to decrypt, unless exact knowledge of the keys is granted.

To validate this, a plaintext image is first encrypted and decrypted, using the correct key values given in (15) the proposed Algorithm 1. Then, the encrypted image is decrypted using the same keys, where one key value is changed each time.

The simulation results are shown in Fig. 15 for the peppers plaintext image. It is clear than in all cases, the decryption using the wrong key values fails. Thus, the encryption design is sensitive with respect to all the key values used.

5.8 Execution Time

In the present section, the execution time of the proposed method is discussed. The results are shown in Table 9. Along with the total execution time, the time required for the different steps of the encryption process is presented, since decomposing the algorithm into its parts enables the discussion on reducing the execution time. The times presented are obtained using a PC with the following specifications:

Operational system	Windows 10 Home (64bit)
CPU	Intel(R) Core(TM) i5-8250U CPU @ 1.60G Hz 1.80 GHz
RAM	8.00 GB DDR4
MATLAB version	R2018a

Note that the process of displaying the encrypted image is not calculated in the total execution time. Clearly, most of the execution time is spent on shuffling the image bits, which was expected. The execution time is where the limitation of the proposed method is identified, as in most recent works, the encryption time ranges from less than a second for greyscale images, to a couple of seconds for coloured images [18, 29, 49, 51, 53, 55]. Thus, in future works, the problem of reducing the execution time of this or similar methods should be addressed.

One potential approach that could reduce the time cost for the row and column level shuffles is to utilize parallel processing when performing the shuffling of different levels. More explicitly, suppose that the image to be encrypted has M rows. In our approach, the rows or columns are shuffled one at a time, as seen next

$$1 \to 2 \to \cdots \to \frac{M(M+1)}{2} \to \frac{M(M+1)}{2} + 1 \to \cdots \to M.$$

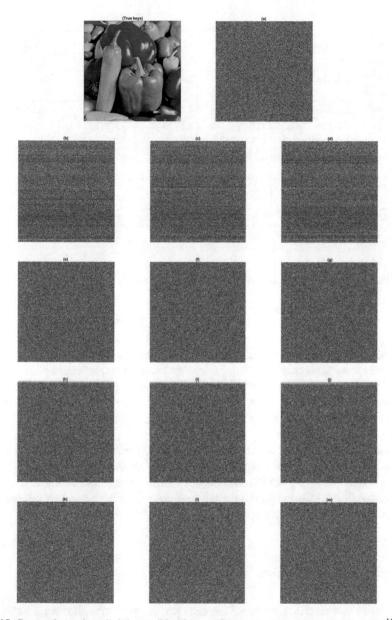

Fig. 15 Decryption using slightly modified keys. **a** Parameter λ is changed to $\lambda + 10^{-16}$, and the rest of the key values are computed based on the new value. **b** Parameter x_0 is changed to $x_0 + 10^{-16}$. **c** Parameter μ_1 is changed to $\mu_1 + 10^{-13}$. **d** Parameter a_1 is changed to $a_1 + 10^{-15}$. **e** Parameter y_0 is changed to $y_0 + 10^{-16}$. **f** Parameter μ_2 is changed to $\mu_2 + 10^{-13}$. **g** Parameter a_2 is changed to $a_2 + 10^{-15}$. **h** Parameter z_0 is changed to $z_0 + 10^{-16}$. **i** Parameter μ_3 is changed to $\mu_3 + 10^{-13}$. **j** Parameter a_3 is changed to $a_3 + 10^{-15}$. **k** Parameter v_0 is changed to $v_0 + 10^{-16}$. **l** Parameter μ_4 is changed to $\mu_4 + 10^{-13}$. **m** Parameter a_4 is changed to $a_4 + 10^{-15}$

Table 9 Time analysis (in seconds) of encryption method for different plaintext images

Image name	Peppers	Airplane	Baboon	Earth	Shades
Dimensions	512 × 512	512 × 512	512 × 512	512 × 512	512 × 512
Arrange bits in 3D matrix	2.4531	2.6563 3	2.5313	2.5938	2.4688
Row shuffle	2.2969	2.2969	2.3281	2.3438	2.3125
Column shuffle	2.25	2.2344	2.1875	2.2813	2.2656
Pixel level shuffle	0.1719	0.1563	0.1719	0.1719	0.1875
XOR and reshape	1.1875	1.2031	1.2813	1.2031	1.2188
Total time	8.35938	8.5469	8.5	8.5938	8.4531

An alternative approach is to perform the shuffling as

$$1 \to 2 \to \cdots \to \frac{M(M+1)}{2} \text{ and } \frac{M(M+1)}{2} + 1 \leftarrow \cdots \leftarrow M - 1 \leftarrow M$$

Clearly, the two sides of the shuffling are disjoint and thus the two processes can be performed simultaneously. The drawback is that a different initialization is required for the "backward" shuffling process as well. A similar approach can be utilized for the columns and the bit level shuffling as well. Furthermore, based on the number of threads available, the number of breaks can increase, which would lead to a further decrease in execution time.

Another alternation that could reduce the execution time is to perform the shuffling process only on half of the image bit levels. It has already been discussed that the fifth to eighth bit levels contain the majority of the image information. This way the computational effort required for the row, column and bit level shuffling of the $M \times N \times 4$ matrix is greatly reduced from the original $M \times N \times 8$ case.

5.9 Overall Evaluation

Considering the results from all of the tests performed above, it can be concluded that the encrypted image is secure with respect to all types of analysis, like histogram analysis, correlation analysis, entropy measure, differential attacks, cropping attacks and noise interference, as well as brute force attacks. Performing only the shuffling procedure on the other hand may yield secure performance with respect to some measures like entropy, correlation, and brute force, but is vulnerable to chosen plaintext attacks, and also histogram analysis. Thus, both steps need to be implemented to have a secure design.

6 Graphical User Interface

In this section, the proposed encryption and decryption method is implemented with a Graphical User Interface (GUI). The reason behind this construction is twofold. Initially, when proposing an encryption and decryption scheme, it is of utmost importance to provide a simple and clear framework, in which anyone can utilize the proposed method, without requiring any familiarity with the theoretical part of the method, or do the implementation method from scratch. Furthermore, considering the fact that such encryption methods can potentially have a hardware implementation, the GUI provides a clear view on what the requirements and functions of such an implementation are.

Moving on to the presentation of the GUI, the interface can be seen in Fig. 16(left). Pressing the "Load Image" button opens a window to the file manager where the user can select the image file to be encrypted or decrypted. A preview of the selected image can be seen on the upper right part, under "Original Image".

To encrypt the loaded image, the "Encrypt and Save Key" button can be pressed. This initializes the encryption process. Once this process is finished, the user is asked to select the destination where the keys required for the decryption are saved in .txt format. Once this is chosen, a preview of the encrypted image is shown on the bottom right side, under the "Resulting Image". Once the encryption is complete, the user can press the "Save Result" button, which allows the user to save the encrypted image. One important notice here is that the encrypted image should not be saved in ".jpg" form, since this leads to round off errors that do not allow for the image to be decrypted.

For the case of decryption, the encrypted image is loaded, and pressing the "Load Decrypt Key" button opens a window where the user selects the .txt file containing the decryption keys. Again, loading the image will cause its preview to appear under "Original Image". Having selected the image file and the keys, pressing the "Decrypt" button will initiate the decryption process. Once finished, a preview of the decrypted image is shown under "Resulting Image" and again the user can save the decrypted image through the "Save Image" button.

In the example shown in Fig. 16, the encryption and decryption of the "peppers" image is considered.

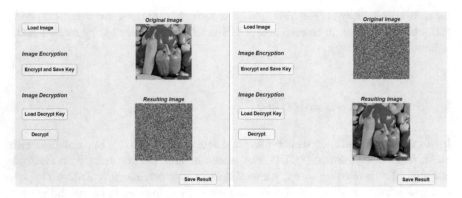

Fig. 16 Utilizing the GUI for the encryption (left) and decryption (right) of the "peppers" image

7 Conclusions

This work presented an image encryption procedure based on bit level shuffling. First, a generalized version of the chaotic map proposed in [8] was developed and its dynamical behaviour was studied through computation of its bifurcation diagrams and Lyapunov exponent diagrams. The map showcased large regions of chaotic behaviour for high parameter values, which makes it suitable for encryption.

Based on the proposed map a secure pseudo-random bit generator was developed. Then, an image encryption system was designed. The system is based on first decomposing an image into its pixel levels, and then performing a shuffling of the bits with respect to each row, column, and bit level. This is followed by an XOR operation, that yields the encrypted image. The resulting image was shown to be random and secure with respect to a collection of different tests and measures. Finally, the design was implemented in a GUI, for ease of use by any interested individual.

Extensions of this work will consider the application to colour images and the combination of this technique with DNA coding. Moreover, steps will be taken to improve the limitations of this work, which is mainly its execution time, by considering permutation only among the most significant bit levels, and also the use of compressed sensing techniques, to reduce file size before encryption. Also, problems like watermarking are of interest.

Acknowledgements The authors would like to thank the anonymous reviewers for their comments that contributed to the improvement of the manuscript's quality.

Appendix

Algorithm 1 Image Encryption

Input: A $M \times N$ grayscale image $\mathcal{A} = [a_{ij}]$, $i = 1, ..., M$, $j = 1, ..., N$.
Four chaotic maps of the form (2) with parameters x_0, μ_1, a_1, $,y_0$, μ_2, a_2, z_0, μ_3, a_3, w_0, μ_4, a_4, computed in Step 1.
Output: An encrypted $M \times N$ grayscale image $\mathcal{E}$.

Step 1.   Compute the parameters of the chaotic maps as

$$\lambda = mod \left(\frac{\sum_{i=1}^{M} \sum_{j=1}^{N} a_{ij}}{M \cdot N} + \text{Ent}(A), 1 \right) \qquad (15a)$$

$$x_0 = \lambda, \qquad\qquad \mu_1 = 900 + \lambda, \qquad\qquad a_1 = 2\pi\lambda \qquad (15b)$$

$$y_0 = mod(10^6\lambda, 1), \qquad \mu_2 = 901 + y_0, \qquad a_2 = 2\pi y_0 \qquad (15c)$$

$$z_0 = mod(10^9\lambda, 1), \qquad \mu_3 = 902 + z_0, \qquad a_3 = 2\pi z_0 \qquad (15d)$$

$$v_0 = \cos(\lambda), \qquad\qquad \mu_4 = 905 + v_0, \qquad a_4 = 2\pi v_0 \qquad (15e)$$

where λ is dependent on the average pixel value of image A, as well as its entropy $\text{Ent}(A)$.

Step 2.   Transform the $M \times N$ image $\mathcal{A}$ into its binary 3D matrix representation $\mathcal{A}_{bin}$ of dimension $M \times N \times 8$.

Step 3.   Perform row-level shuffling of the 3D binary matrix $\mathcal{A}_{bin}$. For each individual row $i = 1, ..., M$, consider the $N \times 8$ binary matrix $\mathcal{R}_i = A_{bin}(i, :, :)$, that is extracted from A_{bin} by taking all the columns set at row i, in all pixel levels. Then, for each $\mathcal{R}_i$ perform the following:

    Step 3.a.   Starting from x_0 for the first row matrix $i = 1$, iterate the chaotic map x_k and compute $p_\tau^i = \lfloor N|x_k| \rfloor + 1$, until N distinct integers are generated in the interval $[1, N]$, that is $\{p_{\tau,1}^i, ..., p_{\tau,N}^i\}$. These indexes denote the row permutations for the 2D matrix $\mathcal{R}_i$. So the p_1^i row of the matrix $\mathcal{R}_i$ is moved to the first row, the p_2^i row is moved to the second row and so are the rest. This procedure is equivalent to the left multiplication $L_i \mathcal{R}_i$, where L_i is an invertible matrix where each row n has zero entries, and 1 in the position $p_{\tau,n}^i$.

    Step 3.b.   Similarly, setting as x_0 the value of the last iteration of the map in the previous step, compute $q_\tau^i = \lfloor 8|x_k| \rfloor + 1$, until 8 distinct integers are generated in the interval $[1, 8]$, that is $\{q_{\tau,1}^i, ..., q_{\tau,8}^i\}$. These indexes denote the column permutations for the 2D matrix $\mathcal{R}_i$. So the q_1^i column of the matrix $\mathcal{R}_i$ is moved to the first column and so on. This procedure is equivalent to the right multiplication $\mathcal{R}_i Q_i$, where Q_i is an invertible matrix where each column n has zero entries, and 1 in the position $q_{\tau,n}^i$.

    Step 3.c.   Repeat this procedure for all matrices $\mathcal{R}_i$, $i = 2, ..., M$. For each individual matrix, the initial value of the chaotic map x_0 is taken as the value of the final iteration from the previous step.

    The resulting shuffled $M \times N \times 8$ matrix is the concatenation of the shuffled matrices $L_i \mathcal{R}_i Q_i$, denoted as $\mathcal{R}$.

Step 4.   Perform column-wise shuffling of the 3D binary matrix $\mathcal{R}$. For each individual column $j = 1, ..., N$, consider the $M \times 8$ binary matrix $C_j = \mathcal{R}(:, j, :)$, that is extracted from $\mathcal{R}$ by taking all the rows set at column j, in all pixel levels. Then for each C_j perform the following:

Algorithm Image Encryption (Continued.)

Step 4. (Cont.)

Step 4.a. Starting from y_0 for the first column $j = 1$, iterate the chaotic map y_k and compute $p_c^j = \lfloor M|y_k| \rfloor + 1$, until M distinct integers are generated in the interval $[1, M]$, that is $\{p_{c,1}^j, ..., p_{c,M}^j\}$. These indexes denote the row permutations for the 2D matrix C_j, which are performed similarly to Step 3a.

Step 4.b. Similarly, setting as y_0 the value of the last iteration of the map in the previous step, compute $q_c^i = \lfloor 8|y_k| \rfloor + 1$, until 8 distinct integers are generated in the interval $[1, 8]$, that is $\{q_{c,1}^j, ..., q_{c,8}^j\}$. These indexes denote the column permutations for the 2D matrix C_j, which are performed similarly to Step 3b.

Step 4.c. Repeat this procedure for all matrices C_j, $j = 2, ..., M$. For each individual matrix, the initial value of the chaotic map y_0 is taken as the value of the final iteration from the previous step.

The resulting column shuffled $M \times N \times 8$ matrix is the concatenation of the shuffled matrices $L_j C_j Q_j$, denoted as C.

Step 5. Perform bit level shuffling of the 3D binary matrix C. For each individual bit level $w = 1, ..., 8$, consider the $M \times N$ binary matrix $\mathcal{P}_w = C(:, :, w)$, that is extracted from C by taking all the rows and columns set at bit level w. Then for each $\mathcal{P}_w$ perform the following:

Step 5.a. Starting from z_0 for the first bit level $w = 1$, iterate the chaotic map z_k and compute $p_p^w = \lfloor M|z_k| \rfloor + 1$, until M distinct integers are generated in the interval $[1, M]$, that is $\{p_{p,1}^w, ..., p_{p,M}^w\}$. These indexes denote the row permutations for the 2D matrix $\mathcal{P}_w$, which are performed similarly to Steps 3a. and 4a.

Step 5.b. Similarly, setting as z_0 the value of the last iteration of the map in the previous step, compute $q_p^w = \lfloor N|z_k| \rfloor + 1$, until N distinct integers are generated in the interval $[1, N]$, that is $\{q_{p,1}^w, ..., q_{p,N}^w\}$. These indexes denote the column permutations for the 2D matrix $\mathcal{P}_w$, which are performed similarly to Steps 3b and 4b.

Step 5.c. Repeat this procedure for all matrices $\mathcal{P}_w$, $w = 2, ..., 8$. For each individual matrix, the initial value of the chaotic map z_0 is taken as the value of the final iteration from the previous step.

The resulting pixel shuffled $M \times N \times 8$ matrix is the concatenation of the shuffled matrices $L_w \mathcal{P}_w Q_w$, denoted as $\mathcal{P}$.

Step 6. After all three shuffling levels are completed, the resulting matrix $\mathcal{P}$ is reshaped into a vector of length $M \cdot N \cdot 8$ and is $XORed$ using a bitstream of the same length, generated from the fourth chaotic map with key values v_0, μ_4, a_4. The resulting bitstream is denoted $\mathcal{E}_{\text{stream}}$.

Step 7. The bitstream $\mathcal{E}_{\text{stream}}$ is reshaped into an $M \times N \times 8$ matrix $\mathcal{E}_{bin}$, which generates the encrypted image $\mathcal{E}$.

References

1. Tang, X., Mandal, S.: Encrypted physical layer communications using synchronized hyperchaotic maps. IEEE Access **9**, 13286–13303 (2021)
2. Ali, Z., Imran, M., Alsulaiman, M., Shoaib, M., Ullah, S.: Chaos-based robust method of zero-watermarking for medical signals. Fut. Gener. Comput. Syst. **88**, 400–412 (2018)

3. Teh, J.S., Tan, K., Alawida, M.: A chaos-based keyed hash function based on fixed point representation. Cluster Comput. **22**(2), 649–660 (2019)
4. Liu, L., Miao, S., Cheng, M., Gao, X.: A pseudorandom bit generator based on new multi-delayed chebyshev map. Inf. Process. Lett. **116**(11), 674–681 (2016)
5. Abdelfatah, R.I., Nasr, M.E., Alsharqawy, M.A.: Encryption for multimedia based on chaotic map: several scenarios. Multimed. Tools Appl. **79**(27), 19717–19738 (2020)
6. Kumar, M., Saxena, A., Vuppala, S.S.: A survey on chaos based image encryption techniques. In: Multimedia Security Using Chaotic Maps: principles and Methodologies, pp. 1–26. Springer (2020)
7. Hua, Z., Zhou, B., Zhou, Y.: Sine chaotification model for enhancing chaos and its hardware implementation. IEEE Trans. Ind. Electron. **66**(2), 1273–1284 (2018)
8. Talhaoui, M.Z., Wang, X., Midoun, M.A.: A new one-dimensional cosine polynomial chaotic map and its use in image encryption. In: The Visual Computer, pp. 1–11 (2020)
9. Liu, L., Miao, S.: A new simple one-dimensional chaotic map and its application for image encryption. Multimed. Tools Appl. **77**(16), 21445–21462 (2018)
10. Zhou, Y., Bao, L., Chen, C.P.: A new 1d chaotic system for image encryption. Signal Process. **97**, 172–182 (2014)
11. Murillo-Escobar, M.A., Cruz-Hernández, C., Cardoza-Avendaño, L., Méndez-Ramírez, R.: A novel pseudorandom number generator based on pseudorandomly enhanced logistic map. Nonlinear Dyn. **87**(1), 407–425 (2017)
12. Alawida, M., Samsudin, A., Teh, J.S.: Enhanced digital chaotic maps based on bit reversal with applications in random bit generators. Inf. Sci. **512**, 1155–1169 (2020)
13. Tutueva, A.V., Nepomuceno, E.G., Karimov, A.I., Andreev, V.S., Butusov, D.N.: Adaptive chaotic maps and their application to pseudo-random numbers generation. Chaos Solitons Fractals **133**, 109615 (2020)
14. Sathya, K., Premalatha, J., Rajasekar, J.: Investigation of strength and security of pseudo random number generators. In: IOP Conference Series: materials Science and Engineering, vol. 1055, pp. 012076. IOP Publishing (2021)
15. Ayubi, P., Setayeshi, S., Rahmani, A.M.: Deterministic chaos game: a new fractal based pseudo-random number generator and its cryptographic application. J. Inf. Secur. Appl. **52**, 102472 (2020)
16. Belazi, A., Talha, M., Kharbech, S., Xiang, W.: Novel medical image encryption scheme based on chaos and DNA encoding. IEEE Access **7**, 36667–36681 (2019)
17. Zhu, C., Wang, G., Sun, K.: Cryptanalysis and improvement on an image encryption algorithm design using a novel chaos based S-box. Symmetry **10**(9), 399 (2018)
18. Zhang, D., Chen, L., Li, T.: Hyper-chaotic color image encryption based on transformed zigzag diffusion and RNA operation. Entropy **23**(3), 361 (2021)
19. Moysis, L., Tutueva, A., Christos, K., Butusov, D.: A chaos based pseudo-random bit generator using multiple digits comparison. Chaos Theory Appl. **2**(2), 58–68 (2020)
20. Moysis, L., Kafetzis, I., Volos, C., Tutueva, A.V., Butusov, D.: Application of a hyperbolic tangent chaotic map to random bit generation and image encryption. In: 2021 IEEE Conference of Russian Young Researchers in Electrical and Electronic Engineering (ElConRus), pp. 559–565. IEEE (2021)
21. Gao, T., Chen, Z.: A new image encryption algorithm based on hyper-chaos. Phys. Lett. A **372**(4), 394–400 (2008)
22. Irani, B.Y., Ayubi, P., Jabalkandi, F.A., Valandar, M.Y., Barani, M.J.: Digital image scrambling based on a new one-dimensional coupled sine map. Nonlinear Dyn. **97**(4), 2693–2721 (2019)
23. Li, Z., Peng, C., Tan, W., Li, L.: A novel chaos-based color image encryption scheme using bit-level permutation. Symmetry **12**(9), 1497 (2020)
24. Ge, R., Yang, G., Jiasong, W., Chen, Y., Coatrieux, G., Luo, L.: A novel chaos-based symmetric image encryption using bit-pair level process. IEEE Access **7**, 99470–99480 (2019)
25. Liu, H., Wang, X.: Color image encryption using spatial bit-level permutation and high-dimension chaotic system. Opt. Commun. **284**(16–17), 3895–3903 (2011)

26. Teng, L., Wang, X.: A bit-level image encryption algorithm based on spatiotemporal chaotic system and self-adaptive. Opt. Commun. **285**(20), 4048–4054 (2012)
27. Kar, M., Mandal, M.K., Nandi, D., Kumar, A., Banik, S.: Bit-plane encrypted image cryptosystem using chaotic, quadratic, and cubic maps. IETE Tech. Rev. **33**(6), 651–661 (2016)
28. Raza, S.F., Satpute, V.: A novel bit permutation-based image encryption algorithm. Nonlinear Dyn. **95**(2), 859–873 (2019)
29. Shahna, K.U., Mohamed, A.: A novel image encryption scheme using both pixel level and bit level permutation with chaotic map. Appl. Soft Comput. **90**, 106162 (2020)
30. Murillo-Escobar, M.A., Meranza-Castillón, M.O., López-Gutiérrez, R.M., Cruz-Hernández, C.: A chaotic encryption algorithm for image privacy based on two pseudorandomly enhanced logistic maps. In: Multimedia Security Using Chaotic Maps: principles and Methodologies, pp. 111–136. Springer (2020)
31. Zhou, Y., Bao, L., Chen, C.L.P.: Image encryption using a new parametric switching chaotic system. Signal Process. **93**(11), 3039–3052 (2013)
32. Wang, X., Liu, L., Zhang, Y.: A novel chaotic block image encryption algorithm based on dynamic random growth technique. Opt. Lasers Eng. **66**, 10–18 (2015)
33. Hsiao, H.-I., Lee, J.: Fingerprint image cryptography based on multiple chaotic systems. Signal Process. **113**, 169–181 (2015)
34. Kari, A.P., Navin, A.H., Bidgoli, A.M., Mirnia, M.: A new image encryption scheme based on hybrid chaotic maps. Multimed. Tools Appl. **80**(2), 2753–2772 (2021)
35. Nepomuceno, E.G., Nardo, L.G., Arias-Garcia, J., Butusov, D.N., Tutueva, A.: Image encryption based on the pseudo-orbits from 1D chaotic map. Chaos: Interdiscip. J. Nonlinear Sci. **29**(6), 061101 (2019)
36. Niu, Y., Zhang, X.: A novel plaintext-related image encryption scheme based on chaotic system and pixel permutation. IEEE Access **8**, 22082–22093 (2020)
37. Zhou, S., He, P., Kasabov, N.: A dynamic DNA color image encryption method based on SHA-512. Entropy **22**(10), 1091 (2020)
38. Gopalakrishnan, T., Ramakrishnan, S.: Chaotic image encryption with hash keying as key generator. IETE J. Res. **63**(2), 172–187 (2017)
39. Zhu, S., Zhu, C., Wang, W.: A new image encryption algorithm based on chaos and secure hash SHA-256. Entropy **20**(9), 716 (2018)
40. Sun, C., Wang, E., Zhao, B.: Image encryption scheme with compressed sensing based on a new six-dimensional non-degenerate discrete hyperchaotic system and plaintext-related scrambling. Entropy **23**(3), 291 (2021)
41. Alvarez, G., Li, S.: Some basic cryptographic requirements for chaos-based cryptosystems. Int. J. Bifurc. Chaos **16**(08), 2129–2151 (2006)
42. Rukhin, A., Soto, J., Nechvatal, J., Smid, M., Barker, E.: A statistical test suite for random and pseudorandom number generators for cryptographic applications. Technical report, Booz-Allen and Hamilton Inc Mclean Va (2001)
43. Safack, N., Iliyasu, A.M., De Dieu, N.J., Zeric, N.T., Kengne, J., Abd-El-Atty, B., Belazi, A., Abd EL-Latif, A.A.: A memristive RLC oscillator dynamics applied to image encryption. J. Inf. Secur. Appl. **61**, 102944 (2021)
44. Abdelfatah, R.I.: Secure image transmission using chaotic-enhanced elliptic curve cryptography. IEEE Access **8**, 3875–3890 (2019)
45. El-Latif, A.A.A., Abd-El-Atty, B., Belazi, A., Iliyasu, A.M.: Efficient chaos-based substitution-box and its application to image encryption. Electronics **10**(12), 1392 (2021)
46. Wu, Y., Zhou, Y., Saveriades, G., Agaian, S., Noonan, J.P., Natarajan, P.: Local shannon entropy measure with statistical tests for image randomness. Inf. Sci. **222**, 323–342 (2013)
47. Vaidyanathan, S., Sambas, A., Abd-El-Atty, B., Abd El-Latif, A.A., Tlelo-Cuautle, E., Guillén-Fernández, O., Mamat, M., Mohamed, M.A., Alçin, M., Tuna, M., et al.: A 5-D multi-stable hyperchaotic two-disk dynamo system with no equilibrium point: circuit design. FPGA realization and applications to trngs and image encryption. IEEE Access (2021)
48. Chen, J.-X., Zhu, Z.-L., Chong, F., Zhang, L.-B., Zhang, Y.: An efficient image encryption scheme using lookup table-based confusion and diffusion. Nonlinear Dyn. **81**(3), 1151–1166 (2015)

49. Sambas, A., Vaidyanathan, S., Tlelo-Cuautle, E., Abd-El-Atty, B., Abd El-Latif, A.A., Guillén-Fernández, O., Hidayat, Y., Gundara, G., et al.: A 3-D multi-stable system with a peanut-shaped equilibrium curve: circuit design, FPGA realization, and an application to image encryption. IEEE Access **8**, 137116–137132 (2020)
50. Tsafack, N., Sankar, S., Abd-El-Atty, B., Kengne, J., Jithin, K.C., Belazi, A., Mehmood, I., Bashir, A.K., Song, O.Y., Abd El-Latif, A.A.: A new chaotic map with dynamic analysis and encryption application in internet of health things. IEEE Access **8**, 137731–137744 (2020)
51. Alanezi, A., Abd-El-Atty, B., Kolivand, H., El-Latif, A., Ahmed, A., El-Rahiem, A., Sankar, S., Khalifa, H., et al.: Securing digital images through simple permutation-substitution mechanism in cloud-based smart city environment. Secur. Commun. Netw. **2021** (2021)
52. Yao, L., Yuan, C., Qiang, J., Feng, S., Nie, S.: Asymmetric color image encryption based on singular value decomposition. Opt. Lasers Eng. **89**, 80–87 (2017)
53. Hua, Z., Zhou, Y., Pun, C.M., Chen, C.P.: 2D sine logistic modulation map for image encryption. Inf. Sci. **297**, 80–94 (2015)
54. Zhu, C., Wang, G., Sun, K.: Improved cryptanalysis and enhancements of an image encryption scheme using combined 1D chaotic maps. Entropy **20**(11), 843 (2018)
55. Nestor, T., De Dieu, N.J., Jacques, K., Yves, E.J., Iliyasu, A.M., El-Latif, A.: A multidimensional hyperjerk oscillator: dynamics analysis, analogue and embedded systems implementation, and its application as a cryptosystem. Sensors **20**(1), 83 (2020)

Adaptive Chaotic Maps in Cryptography Applications

Aleksandra Tutueva, Erivelton G. Nepomuceno, Lazaros Moysis, Christos Volos, and Denis Butusov

Abstract Chaotic cryptography is a promising area for the safe and fast transmission, processing, and storage of data. However, many developed chaos-based cryptographic primitives do not meet the size and composition of the keyspace and computational complexity. Another common problem of such algorithms is dynamic degradation caused by computer simulation with finite data representation and rounding of results of arithmetic operations. The known approaches to solving these problems are not universal, and it is difficult to extend them to many chaotic systems. This chapter describes discrete maps with adaptive symmetry, making it possible to overcome several disadvantages of existing chaos-based cryptographic algorithms simultaneously. The property of adaptive symmetry allows stretching, compressing, and rotating the phase space of such maps without significantly changing the bifurcation properties. Therefore, the synthesis of one-way piecewise functions based on adaptive maps with different symmetry coefficients supposes flexible control of the keyspace size and avoidance of dynamic degradation due to the embedded technique of perturbing the chaotic trajectory.

A. Tutueva (✉) · D. Butusov
Youth Research Institute, Saint-Petersburg Electrotechnical University "LETI", 5, Professora
Popova st., 197376 Saint Petersburg, Russia
e-mail: avtutueva@etu.ru

D. Butusov
e-mail: dnbutusov@etu.ru

E. G. Nepomuceno
Control and Modelling Group (GCOM), Department of Electrical Engineering,
Federal University of São João del-Rei, São João del-Rei, MG 36307-352, Brazil
e-mail: nepomuceno@ufsj.edu.br

L. Moysis · C. Volos
Laboratory of Nonlinear Systems - Circuits & Complexity (LaNSCom), Physics Department,
Aristotle University of Thessaloniki, Thessaloniki, Greece
e-mail: lmousis@physics.auth.gr

C. Volos
e-mail: volos@physics.auth.gr

1   Introduction

Information security is one of the widely investigated fields of deterministic chaos practical applications [1–6]. An analysis of data from *Scopus* and *Web of Science* indexing systems shows that over the past two decades, more than five and a half thousand papers have been published on chaos-based cryptography (Fig. 1). Moreover, there is a trend towards increasing the number of studies carried out in this research area. In 2020, the publications indexed in *Scopus* and devoted to the development of chaos-based cryptographic primitives were about 20% of all articles on chaotic systems (Fig. 2).

The high research activity in this field has led to developing the presentation requirements of theoretical and experimental results. In 2006, Alvarez and Li proposed seventeen rules that should guide the description, implementation, keyspace generation, and security analysis of new chaos-based cryptosystems [7]. The motivation for developing such a framework was that many of the reported approaches do not possess several fundamentally essential features to secure data storage, processing, and transmission. By 2021, this article has been cited more than one thousand times, which means that the scientific community has generally accepted the proposed rules as the standard for chaos-based cryptosystems. Although more than

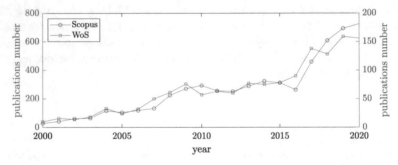

Fig. 1   The number of publications on chaos-based cryptography in *scopus* and *web of science*

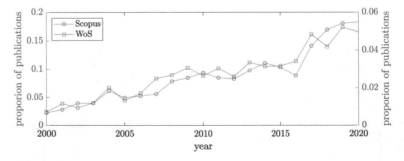

Fig. 2   Proportion of publications on chaotic cryptography out of the total number of articles on chaotic systems in *scopus* and *web of science*

fifteen years have passed since the study by Alvarez and Lee has been published, a universal solution has not yet been proposed that is devoid of all the considered shortcomings of chaos-based cryptosystems. Most of the presented research aims to solve a specific issue, such as expanding the keyspace or improving the statistical properties of ciphertexts.

This chapter considers the concept of recently proposed adaptive chaotic maps and discusses the advantages of such discrete systems for chaos-based cryptography. We summarize the results of our previous research [8–12] and reveal the potential of the adaptive maps family as a possible unified solution of problems raised in [7], including reduction of the degradation of chaotic dynamics, rigorous choice of the keyspace, and flexible control of its size. We explicitly show that the symmetry coefficient as the additional parameter expands the keyspace, making it easier to choose the values that generate chaotic trajectories. In addition, we propose a technique that allows breaking the cycles that arise from the finite data representation and rounding the results of arithmetic operations.

The rest of the chapter is organized as follows. In Sect. 2, we discuss the common issues of chaos-based cryptography primitives and their existing solutions. Section 3 describes the adaptive maps family and gives an example of the adaptive Hénon map. In Sect. 4, we consider the ways that adaptive maps can improve chaos-based cryptosystems. Finally, some conclusions and discussions are given in Sect. 5.

2 Chaos-Based Cryptosystem Problems

All cryptographic requirements described in [7] could be grouped into three sets: implementation rules, the definition of the keyspace, and instructions for the security analysis. In this study, we focus on the first two groups since the properties of the used chaotic system directly affect these aspects of the cryptosystem, in contrast to security, which often depends on attended algorithmic procedures such as mixing and shuffling [13].

2.1 Implementation issues

According to rules [7], the chaos-based cryptosystem should be easy to implement with acceptable cost and speed without loss of security. Moreover, it is required to introduce techniques that reduce the chaotic dynamics degradation arising from the finite precision of data representation in computer simulation of chaotic systems.

Several approaches to overcome dynamics degradation have been recently proposed [14–25]. Some of the described techniques use piecewise linear chaotic maps. Due to several definition domains, such systems do not strictly depend on state variables values obtained in the previous iteration. In this case, the accumulation of the rounding errors of the results of arithmetic operations is slower. Therefore, for piecewise linear maps, the period of a chaotic sequence can often be higher than for other maps of the same dimension. Recently, many encryption algorithms have been proposed based on the one-dimensional piecewise linear chaotic map [14], the tent map [15], the discrete Chebyshev third-order chaotic map [16], and the compound chaotic function switched by the linear feedback shift register (LFSR) [17, 18]. However, all of the mentioned approaches have been described for the specific systems and are difficult to extend to other chaotic maps.

Another possible solution to the abovementioned problem is considered in [19]. The reported approach uses a combination of chaotic systems to reduce the influence of the finite data representation and round off arithmetic operations. The proposed algorithm applies a mixing technique based on the logistic map for image pixels permutation. Then, an encryption algorithm that implements the Chen continuous system is utilized to the result. The proposed method allows obtaining chaotic sequences with an extended period. However, it does not meet the requirement to minimize computational costs, which is significantly higher when implementing cryptographic primitives based on continuous chaotic systems than discrete maps. The algorithm considered in [20] has a similar problem. In this method, one should simultaneously simulate two different chaotic maps to improve the degree of randomness, which positively affects the period length of the chaotic trajectory. Compared with study [19], this approach provides for parallel execution but requires additional research on the overall dynamics of the obtained map. A simultaneous change in nonlinearity parameters can lead to unacceptable multiparametric bifurcations or cause periodic oscillations [21]. A similar effect may occur while overcoming dynamic degradation throughout perturbation of the parameter, or chaotic trajectory [22–25]. Thus, there is still no universal and robust solution for increasing the period of chaotic sequences in computer simulations.

2.2 Keyspace definition problems

In chaos-based cryptography, the keyspace of encryption algorithms, hash functions, or other primitives often includes initial values and bifurcation parameters [26]. Therefore, for the chaotic systems used, the following rules are established [7]:

1. The key should be precisely defined.
2. The keyspace, from which valid keys are to be chosen, should be precisely specified and avoid nonchaotic regions.
3. The useful chaotic region should be discretized so that the avalanche effect is guaranteed. It means that two ciphertexts encrypted by two slightly different keys should be completely different.
4. To provide sufficient security against brute-force attacks, the keyspace size should be larger than 2^{100}.

Fig. 3 Two-dimensional bifurcation diagram for the Hénon map. Chaotic regions correspond to white color, nonchaotic regions are marked black

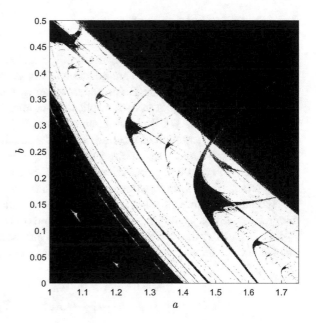

The third requirement is usually provided by the topological mixing property of the chaotic trajectories [27]. The fulfillment of the rest of the rules directly depends on the chaotic model chosen for implementation in the cryptographic algorithm. To achieve a needed size of the keyspace, chaotic systems with many state variables and several bifurcation parameters are often used [28–34]. Along with the complexity increasing and the possible multiparametric bifurcations, this complicates precisely defining the keyspace. For example, from the two-dimensional bifurcation diagram (Fig. 3) obtained for the Hénon map given by equations

$$x_{n+1} = 1 - ax_n^2 + y_n$$
$$y_{n+1} = bx_n;$$
(1)

with a and b parameters, it is shown that the thorough distinction of non-chaotic and chaotic regions is challenging. Moreover, in the case of continuous chaotic systems, the used discretization method can introduce its nonlinear effects that can distort the bifurcation properties of the prototype system [35]. Thus, to expand the keyspace, the approach based on high-dimensional systems with many nonlinearity parameters always requires rigorous and laborious analysis. The development of models with chaotic dynamics for which the study of bifurcation properties is simpler is of interest.

3 Adaptive Symmetry Concept

The adaptive chaotic maps extend the family of chaotic maps with phase space symmetry [8, 9]. Such systems are generated through the application of geometric integration techniques to the Hamiltonian systems. One of the possible methods for obtaining symmetric schemes implies the composition of two *adjoint* Euler-Cromer methods [36, 37]. Both integrators are applied with half the integration step h. Geometric interpretation of a certain integration method is presented in (Fig. 4a). To obtain adaptive maps, the coefficient 0.5 is replaced by the adaptive coefficient s also called the symmetry coefficient (Fig. 4b). Such a substitution gives a family of discrete systems with rotation control in phase space. In continuous systems, the symmetry coefficient allows to control the stability of such a compositional integration method by varying the value of s from 0 to 1 [38–40].

Let us consider the described approach with the example of the adaptive Hénon map. It was obtained from the symmetric Hénon map [9] given by the equations

$$q_{n+1} = p_n - 0.5(q_n^2 - K)$$
$$p_{n+1} = -q_n - 0.5(q_{n+1}^2 - K) \tag{2}$$

where K is the bifurcation parameter. The value 0.5 in system (2) is replaced by the adaptive coefficient s as follows

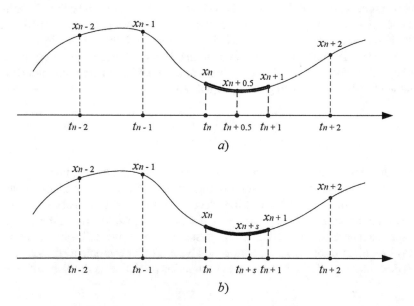

Fig. 4 Geometric interpretation of the adaptive symmetry technique

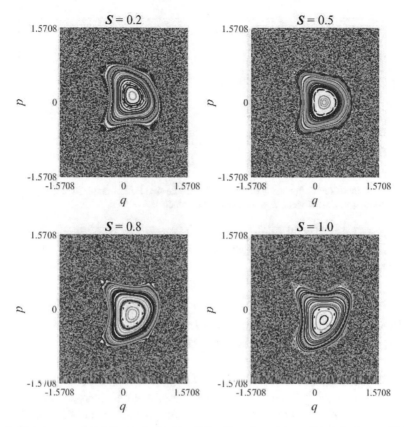

Fig. 5 Phase space of adaptive maps for $K = 0.5$ with various values of the adaptive coefficient

$$q_{n+1} = p_n - s(q_n^2 - K)$$
$$p_{n+1} = -q_n - (1 - s)(q_{n+1}^2 - K)$$

(3)

The system (3) is called the adaptive Hénon map. Figure 5 presents the examples of the phase space corresponding to the various values of the adaptive coefficient. Both state variables are taken modulo 2π. One can see that the case of $s = 0.5$ provides symmetry with respect to $p = 0$. Moreover, a change in the symmetry coefficient leads to phase portrait stretching, compression, and rotation. Thus, one can assume that the adaptive coefficient affects the bifurcation properties of the system to a lesser extent than the parameter K.

To check the abovementioned hypothesis, we plotted the two-dimensional bifurcation diagram (Fig. 6). As one can see, for almost all values of the parameter K, the adaptive Hénon map generates chaotic trajectories. Moreover, it should be noted that for values of the bifurcation parameter exceeding 3, regardless of s, the system exhibits chaotic behavior. The similar phenomena was observed for other previously studied adaptive maps [10–12].

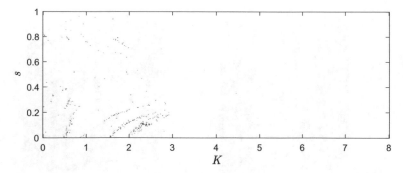

Fig. 6 Two-dimensional bifurcation diagram for the adaptive Hénon map. Chaotic regions correspond to white color, nonchaotic regions are marked black

Let us consider how the adaptive chaotic maps can improve the chaos-based cryptography algorithms.

4 Improving Chaos-Based Cryptographic Algorithms Through the Maps With Adaptive Symmetry

The weak dependence of bifurcation properties on the symmetry coefficient in adaptive maps raises the following advantages for chaos-based cryptography. If we consider the adaptive coefficient as an additional parameter in hash functions or encryption algorithms, it simplifies determining the keyspace. Comparing the two-dimensional bifurcation diagrams for the conventional Hénon map (Fig. 3) and its adaptive modification (Fig. 6), we can conclude that by introducing the restriction $K > 3$ in the second case, we exclude the generation of periodic oscillations. However, the same simple rule cannot be formulated for map (1). Therefore, the first and second requirements from the list mentioned in Sect. 2.2 are easier to fit in the case of adaptive maps while the computational complexity of both chaotic systems is equal.

The dimension of any adaptive map is at least two, while the system equations usually include the symmetry coefficient and the bifurcation parameter. Thus, the keyspace consists of four real values. In the computer simulation with the *double* floating-point data type, these four keys are represented with the precision of 53 bits [41], which yields 2^{212}. It means that the requirement for the keyspace size of the chaotic system used as the foundations of cryptography algorithms is fulfilled [42–44]. Moreover, piecewise functions based on adaptive chaotic maps can be used to increase and flexibly control the keyspace. There are two general approaches to get such maps.

The first idea is to match different values of the adaptive coefficient to several sub-domains of the chaotic map. Let s_1 and s_2 be two values of adaptive symmetry coefficients. The value of s for each iteration is chosen as

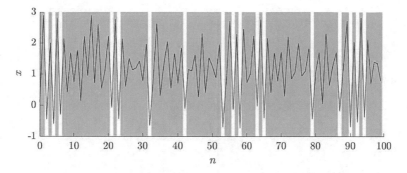

Fig. 7 Time domain for the piecewise adaptive Hénon map

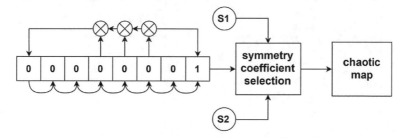

Fig. 8 The adaptive map perturbation process by LFSR

$$s = \begin{cases} s_1, x_i < t \\ s_2, x_i \geqslant t \end{cases} \tag{4}$$

where t is the threshold value. One can choose any value of t that the variable x can take. Fig. 7 shows the switching of the symmetry coefficient value while $t = 0$, $x_0 = 0.1$, $p_0 = -0.1$, $s_1 = 0.5$, $s_2 = 1$. The red color corresponds to the iterations when s was equal to 1. To expand the keyspace size, one can increase the number of subdomains in function (4).

To exclude the dependence of the choice of the symmetry coefficient on the calculated state variable value, one can use the LFSR [12]. According to such a technique, each adaptive coefficient s_1 or s_2 is matched to the LFSR output. Figure 8 represents one of the possible schemes to perturb chaotic trajectories using the proposed approach with LFSR controlled by primitive polynomial $y^8 + y^6 + y^5 + y^4 + 1$.

Using polynomials of a higher degree and encoding different symmetry coefficients with several consecutive bits, it is possible to control the keyspace size. Moreover, switching the coefficient in both considered approaches helps to reduce dynamic degradation. Figure 9 shows the results of estimating the start time of the periodic mode of a chaotic sequence for both initial conditions varied in the interval [0.03125; 0.96875] when simulating with the 16-bit fixed-point data type (8 bit for the word length). For each pair of initial conditions, 500 iterations were calculated with different K. The white color corresponds to the maximum period, i.e., each

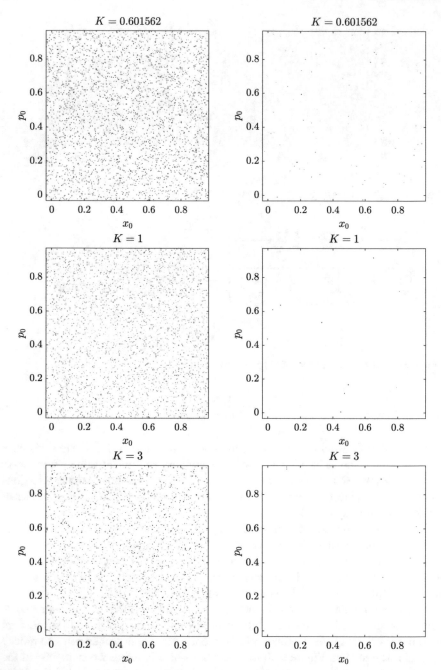

Fig. 9 Estimations of the start time of the periodic mode in the chaotic sequence obtained for different K

sequence value was repeated only once. The left column shows the estimation results for the original adaptive Henon map, and the right column presents the obtained estimations for the perturbation technique based on the LFSR with $s_1 = 0.300781$ and $s_2 = 0.699219$. One can see that the proposed technique is suitable for break out the cycles in chaotic sequences.

5 Conclusion and Discussion

In this study, the application of adaptive chaotic maps to improve chaos-based cryptography was investigated. The theoretical and experimental results show that such systems can be helpful to overcome dynamic degradation and increase the keyspace size, which are common problems of existing chaos-based cryptographic algorithms. Moreover, it can be assumed that adaptive chaotic maps with different symmetry coefficients considered in this study, as well as piecewise linear functions, are less susceptible to attack by phase space reconstruction [45, 46].

In addition to adaptive discrete maps, finite-difference schemes with controllable symmetry obtained from continuous chaotic systems through symmetric integration are of interest for solving information security problems. In paper [38], it was shown that such schemes are suitable foundations for implementing steganographic data transmission through chaotic synchronization. Compared to the traditional approach, implying modulation of the chaotic model parameter, this technique is more secure from the possibility of recognizing the carrier signal. Therefore, investigating new applications of chaotic models with controlled symmetry is a promising direction for further research.

Funding

This study was supported by the grant of the Russian Science Foundation (RSF), project 20-79-10334.

Acknowledgements We thank the anonymous reviewers for their careful reading of our manuscript and their many insightful comments and suggestions.

References

1. Andrievskii, B., Fradkov, A.: Autom. Remote Control **65**(4), 505 (2004)
2. Vaidyanathan, S., Sambas, A., Abd-El-Atty, B., Abd El-Latif, A.A., Tlelo-Cuautle, E., Guillén-Fernández, O., Mamat, M., Mohamed, M.A., Alçin, M., Tuna, M., et al.: IEEE Access (2021)
3. Alanezi, A., Abd-El-Atty, B., Kolivand, H., El-Latif, A., Ahmed, A., El-Rahiem, A., Sankar, S., Khalifa, H.S., et al.: Secur. Commun. Netw. **2021** (2021)

4. El-Latif, A.A.A., Abd-El-Atty, B., Belazi, A., Iliyasu, A.M.: Electronics **10**(12), 1392 (2021)
5. Sambas, A., Vaidyanathan, S., Tlelo-Cuautle, E., Abd-El-Atty, B., Abd El-Latif, A.A., Guillén-Fernández, O., Hidayat, Y., Gundara, G., et al.: IEEE Access **8**, 137116 (2020)
6. Tsafack, N., Sankar, S., Abd-El-Atty, B., Kengne, J., Jithin, K., Belazi, A., Mehmood, I., Bashir, A.K., Song, O.Y., Abd El-Latif, A.A.: IEEE Access **8**, 137731 (2020)
7. Alvarez, G., Li, S.: Int. J. Bifurc. Chaos **16**(08), 2129 (2006)
8. Karimov, A.I., Butusov, D.N., Rybin, V.G., Karimov, T.I.: 2017 XX IEEE International Conference on Soft Computing and Measurements (SCM), pp. 341–344. IEEE (2017)
9. Butusov, D.N., Karimov, A.I., Pyko, N.S., Pyko, S.A., Bogachev, M.I.: Phys. A: Stat. Mech. Appl. **509**, 955 (2018)
10. Tutueva, A.V., Nepomuceno, E.G., Karimov, A.I., Andreev, V.S., Butusov, D.N.: Chaos Solitons Fractals **133**, 109615 (2020)
11. Tutueva, A.V., Karimov, A.I., Moysis, L., Volos, C., Butusov, D.N.: Chaos Solitons Fractals **141**, 110344 (2020)
12. Tutueva, A., Pesterev, D., Karimov, A., Butusov, D., Ostrovskii, V.: 2019 25th Conference of Open Innovations Association (FRUCT), pp. 333–338. IEEE (2019)
13. Schneier, B.: Applied Cryptography: protocols, Algorithms, and Source Code in C. Wiley (2007)
14. Wang, X.Y., Wang, X.J.: Int. J. Modern Phys. C **19**(05), 813 (2008)
15. Guanghui, C., Xing, Z., Yanjun, L.: IETE Tech. Rev. **33**(3), 297 (2016)
16. Qiao, Z., El Assad, S., Taralova, I.: AEU-Int. J. Electron. Commun. **121**, 153205 (2020)
17. Tong, X., Liu, Y., Zhang, M., Wang, Z.: 2012 11th International Symposium on Distributed Computing and Applications to Business, Engineering & Science, pp. 285–289. IEEE (2012)
18. Tong, X.J., Zhang, M., Wang, Z., Liu, Y.: Nonlinear Dyn. **78**(3), 2277 (2014)
19. Chen, D., Zhu, Z., Yang, G.: 2008 The 9th International Conference for Young Computer Scientists, pp. 2792–2796. IEEE (2008)
20. Xiang, H., Liu, L.: Multimedia Tools and Applications, pp. 1–28 (2021)
21. Cai, L., Chen, G., Xiao, D.: J. Math. Biol. **67**(2), 185 (2013)
22. Shu-Bo, L., Jing, S., Zheng-Quan, X., Jin-Shuo, L.: Chin. Phys. B **18**(12), 5219 (2009)
23. Wu, Q., Wang, G., Yuan, L.: 2010 International Workshop on Chaos-Fractal Theories and Applications, pp. 159–163. IEEE (2010)
24. Liu, L., Lin, J., Miao, S., Liu, B.: Int. J. Bifurc. Chaos **27**(07), 1750103 (2017)
25. Chen, C., Sun, K., He, S.: Signal Process. **168**, 107340 (2020)
26. Alvarez, G., Amigó, J.M., Arroyo, D., Li, S.: Chaos-Based Cryptography, pp. 257–295. Springer (2011)
27. Kocarev, L.: IEEE Circ. Syst. Mag. **1**(3), 6 (2001)
28. Behnia, S., Akhshani, A., Mahmodi, H., Akhavan, A.: Chaos Solitons Fractals **35**(2), 408 (2008)
29. Farajallah, M., El Assad, S., Deforges, O.: Int. J. Bifurc. Chaos **26**(02), 1650021 (2016)
30. Gafsi, M., Abbassi, N., Hajjaji, M.A., Malek, J., Mtibaa, A.: Sci. Program. **2020** (2020)
31. Som, S., Dutta, S., Singha, R., Kotal, A., Palit, S.: Nonlinear Dyn. **80**(1), 615 (2015)
32. Elgendy, F., Sarhan, A.M., Eltobely, T.E., El-Zoghdy, S.F., El-Sayed, H.S., Faragallah, O.S.: Multimed. Tools Appl. **75**(18), 11529 (2016)
33. Tresor, L.O., Sumbwanyambe, M.: IEEE Access **7**, 103463 (2019)
34. Hsiao, H.I., Lee, J.: Signal Process. **113**, 169 (2015)
35. Butusov, D., Karimov, A., Tutueva, A., Kaplun, D., Nepomuceno, E.G.: Entropy **21**(4), 362 (2019)
36. Butusov, D., Karimov, A., Andreev, V.: 2015 XVIII International Conference on Soft Computing and Measurements (SCM), pp. 78–80. IEEE (2015)
37. Butusov, D.N., Pesterev, D.O., Tutueva, A.V., Kaplun, D.I., Nepomuceno, E.G.: Commun. Nonlinear Sci. Numer. Simul. **92**, 105467 (2021)
38. Karimov, T., Rybin, V., Kolev, G., Rodionova, E., Butusov, D.: Appl. Sci. **11**(8), 3698 (2021)
39. Tutueva, A.V., Karimov, T.I., Andreev, V.S., Zubarev, A.V., Rodionova, E.A., Butusov, D.N.: 2020 Ural Smart Energy Conference (USEC), pp. 143–146. IEEE (2020)

40. Rybin, V., Tutueva, A., Karimov, T., Kolev, G., Butusov, D., Rodionova, E.: 2021 10th Mediterranean Conference on Embedded Computing (MECO), pp. 1–4. IEEE (2021)
41. Kahan, W.: Lect. Notes Status IEEE **754**(94720–1776), 11 (1996)
42. Norouzi, B., Mirzakuchaki, S.: Nonlinear Dyn. **78**(2), 995 (2014)
43. Hu, T., Liu, Y., Gong, L.H., Ouyang, C.J.: Nonlinear Dyn. **87**(1), 51 (2017)
44. Li, C., Lin, D., Feng, B., Lü, J., Hao, F.: IEEE Access **6**, 75834 (2018)
45. Xie, N., Leung, H.: Proceedings of the 2003 International Symposium on Circuits and Systems, 2003. ISCAS'03, vol. 3, pp. III–III. IEEE (2003)
46. Xie, N., Leung, H.: IEEE Trans. Circ. Syst. I: Regul. Pap. **51**(6), 1210 (2004)

Efficient Secure Medical Image Transmission Based on Brownian System

P. Kiran, H. T. Panduranga, and J. Yashwanth

Abstract This article introduces an automated area of interest (ROI) based encryption method for applying security to medical images. The benefit of utilising this strategy is that it is resistant to a variety of attacks, including Median, Wiener, Gaussian, and differential attacks. The suggested technique is divided into three parts: an automatic ROI detection system, a permutation process, and a diffusion process. The position of each pixel in the ROI image that was changed using the Arnold map sequence and the pixel values are manipulated using the Brownian map. The results obtained with the proposed strategy showed that it has a promising performance.

Keywords Segmentation · Encryption · ROI detection · Arnold map · Brownian map

1 Introduction

Medical services and information technology have recently piqued the interest of many people, resulting in a slew of improvements in the medical industry. Telemedicine is a data-centric therapeutic field in which a massive amount of data is created and stored on a distributed cloud by default. Medical image processing is the most common work in telemedicine, where pictures from x-rays, CT scans, and MRIs can be shared to treat and diagnose disorders. At the same time, as patient data migrates for storage and processing in a distributed environment, this simple transfer increases data protection and security issues. In order to provide confidentiality and a secure environment.

Over the last few decades, extensive research has been conducted on features of secure picture transmission and retrieval approaches. Many approaches for

P. Kiran (✉) · J. Yashwanth
Department of ECE, Vidyavardhaka College of Engineering, Mysuru, Karnataka, India

H. T. Panduranga
Department of ECE, Government Polytechnic, Turvekere, Tumkur, Karnataka, India

protecting medical images using chaotic maps in their pixel scrambling and diffusion stages have been developed [1–4]. Despite the fact that these approaches guarantee randomness during encryption, their fixed secret keys can make them subject to classic attacks. To create a keystream that is both safe and efficient. Xingyuan et al. [5] suggested an image encryption technique based on LL-Compound-Chaos and ZigZag-Transformation. The image sensitivity is determined by local pixels, resulting in low global entropy measurements. In addition, many encryption methods based on one-time keys [6–8] are introduced to ensure resistance against cryptanalytic attacks. For encrypting the medical image, Boussif [9] suggested an adaptive block key. By adding random data into the image for the shuffling process, Hua Z et al. [10] suggested high-speed pixel shuffling and adaptive diffusion-based medical image encryption. Because the reversible procedure can result in data loss during the image compression step, it is time consuming. For image encryption, Abd-El-Atty [11] developed a logistic Chebyshev card-based S-box and a pseudo-random number generator. The system does, however, have a lower UACI value than the proposed picture encryption method. The other option is to ignore the ROI during the watermarking process by leaving this section blank and relying on reversible [12–17] or non-reversible [18–21] concentrated data embedding in the area of no interest (RONI). Amin and Abd El-Latif [22] explained amount modified RC5 based chaotic image encryption technique. This method can be used for image encryption by adjusting the structure of both the encryption routine and the key schedule. Belazi et al. [23] proposed new selective encryption scheme based on DWT, AES S-box and chaotic permutation is proposed. The new scheme is composed of six steps: Image decomposition, Block permutation, DWT decomposition, substitution phase, chaotic permutation phase and reconstruction phase. Tsafack et al. [24] proposed an effective cryptosystem for safeguarding medical picture transmission in an Internet of Healthcare Things (IoHT) context. For added protection, separate maps are used. In order to preserve the privacy of patients, Abd El-Latif et al. [25] presented a new encryption algorithm for privacy-preserving Internet of Things-based healthcare systems. Controlled alternative quantum walks are used in the encryption and decryption procedures.

Novelty of the proposed work as follows.

- A lossless encryption and decryption algorithm for medical image ROI is proposed. The lossless decryption of medical images is crucial for medical staff and patients.
- The hiding of the position information of the ROI in the encryption process is realized, and the information leakage caused by the embedding or transmission of the position information of the ROI is avoided.

Rest of the paper organised as follows. Section 2 explains the ROI detection system and different chaotic system that is used in the proposed work. Section 3 gives the proposed works architecture. Section 4 give the experimental analysis of proposed work and Sect. 5 conclude the work.

2 ROI Detection Method

The proposed ROI identification mechanism is depicted in Fig. 1. Each stage is explained in detail in the following sections.

2.1 Morphological Reconstruction

In image processing, for representing and describing features such as boundaries and skeletons, the morphological operation has become one of the most important mathematical tools. Structuring element, dilatation, and erosion are the three essential notions that all morphological procedures are built on. The structural element is a 0 and 1 matrix with various forms and sizes that regulates the degree of growth or contraction.

Erosion is a method of reducing the size of a thing (see Fig. 2b). Each object pixel that comes into contact with a backdrop pixel is turned to a background pixel in this operation. Dilation is a two-step erosion process. As demonstrated in Fig. 2c, this method is utilised to thicken objects. Each background pixel connected with the object pixel is transformed to the object pixel after dilation. The opening operation is a secondary morphological process that is accomplished through erosion and

Fig. 1 ROI automatic detection flowchart

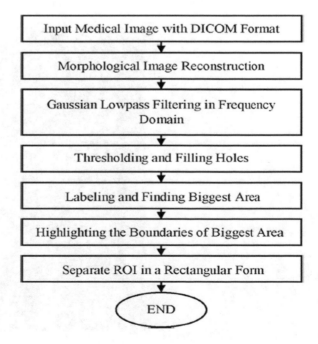

Fig. 2 Green colour
highlighted ROI's boundary

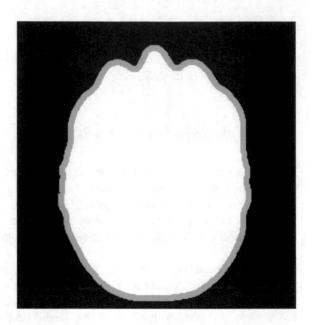

dilatation. Another secondary step is closure, which is accomplished via dilatation
followed by closure erosion (Fig. 3).

Fig. 3 Segmented ROI in a
rectangle

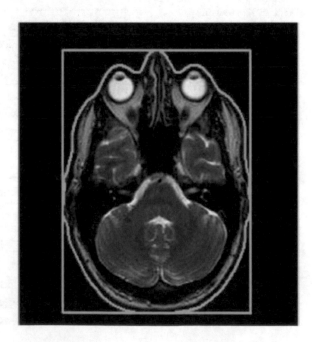

2.2 Arnold Map

The Arnold chaotic map, which is a generalised two-dimensional Arnold chaotic map [26], is

$$\begin{bmatrix} g' \\ h' \end{bmatrix} = \begin{bmatrix} 1 & p \\ q & 1+pq \end{bmatrix} \begin{bmatrix} g \\ h \end{bmatrix} \mathrm{mod}\, N \tag{1}$$

N is the image size, and p and q are the control parameters. (g',h') and (g, h) are random coordinate and original coordinate respectively. The Arnold chaotic map is a tool that is used to shuffle the input medical image in the permutation process.

2.3 Brownian Motion (BM)

Brownian motion is the spontaneous movement of particles in a liquid or gaseous medium caused by interactions between fast moving atoms and molecules. The evolution of particles in three primary directions (here X and Y) is mathematically stated as

$$X = r\, \mathrm{sin} a\, \mathrm{cos} b,\, Y = r\, \mathrm{sin} a\, \mathrm{sin} b \tag{2a}$$

$$a = u_i \times 2 \times \pi,\, b = v_i \times \pi \tag{2b}$$

When enough information about the direction of travel of the particles, i.e., particles moving in three directions, the state of the Brownian particle can be detected (X and Y). The overall.

number of particles (np) participating in the zig-zag motion, as well as the number of impulses per change in track associated with a zig-zag motion, all contribute to the specific time duration (tp). The pseudo-random function is used to select the direction of particle movement, and the step length is denoted by $r = 2$. The X and Y properties of each Brownian particle's position can be derived in this manner. The Monte Carlo method can be used to create BM. We simulated the Brownian motion of 300 particles, the movement trace of a particle is shown in Fig. 4, and the figures indicate that the motion has strong randomness at different times.

3 Proposed ROI Part Encryption System

Figure 5 shows a block schematic of the proposed ROI-based encryption system. There are three parts to the proposed method: ROI identification, permutation, and diffusion. First, extract the ROI fraction from the medical input image using the ROI

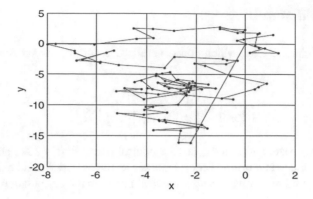

Fig. 4 Single Brownian particle motion path

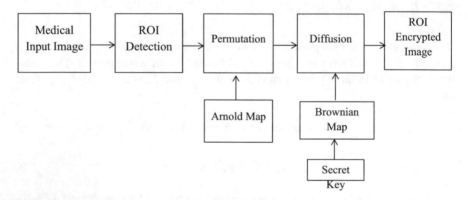

Fig. 5 Block diagram of proposed automatic ROI encryption system

detection method, as explained in Sect. 2. Then the position of each pixel in the ROI image is randomly shuffled using the Arnold cat map. In the case of permutation, the correlation between the neighbouring pixels is reduced. In the diffusion process, a key image is generated with the aid of a Brownian map and a bit-wise XOR operation is carried out between the permuted image and the key image in order to generate an ROI-encrypted image.

The steps involved in the proposed encryption method are presented as an algorithm.

Algorithm

Input: Grayscale image of size ($M \times N$).

Output: Encrypted image of size ($M \times N$).

Begin

Step 1: Read plain image of size ($M \times N$).

Step 2: Apply Morphological operation on the plain image to detect the ROI and RONI (Region of Not Interest) part.

Step 3: Divide the ROI image obtained from Step 2 into (16 × 16) non-overlapping.

Step 4: Apply Arnold cat map for every 16 × 16 blocks obtained in step 3 to reduce the correlation among the adjacent pixels in the block.

Step 5: Iterate Brownian system (M × N) times with initial conditions.

Step 6: Based on digitized sequence Sx, convert one of the sequence from Brownian system.

Integer sequence Ix. It can be represented.

As.

$$Ix \leftarrow Mod(Sx * 10^9, 256)$$

Step 7: rearrange the Ix sequence into matrix of size (M × N) and call it as key image.

Step 8: Diffuse the shuffled block by performing Exclusive-OR (XOR) between.

Values from key image and shuffled block.

Step 9: Finally combined the encrypted ROI block and unencrypted RONI part to produce the ROI encrypted image.

End

Decryption is a reverse process of encryption. The decryption should be performed on the receiving end to extract the original image from the encrypted image transmitted over an unsecured channel. The proposed method used a lossless encryption technique since the reconstructed image obtained after the decryption process is exactly the same as the original image. The encryption steps should be done in reverse order to get the decrypted image. The figure shows the workflow of the decryption process.

4   Performance Analysis of Proposed Scheme

Different parameters should be evaluated to analysis the performance of proposed scheme. The following parameters are involved as follows.

a. **Entropy Analysis**

In an encryption system, entropy is a measure of the degree of randomness. The formula is used to compute the entropy. [27]:

$$H(S) = \sum_{i=0}^{2^M-1} P(\text{si})\log_2 \frac{1}{P(si)} \tag{3}$$

where $P(\text{si})$ is the probability of the ith grey level appearing in an image. For a random image, the ideal entropy value is 8. If it's lower, there's a better likelihood of predictability. The entropy of some example images and their matching cipher images is shown in Table 3.

b.   **Mean Square Error (MSE)**

In general, MSE is calculated by taking the mean of the squared difference between the plain image and the encrypted image. A higher MSE value means more encryption and noise in the plain image. MSE [27] has a mathematical equation, which is provided by.

$$\text{MSE} = \frac{1}{M \times N} \sum_{i=1}^{M} \sum_{j=1}^{N} [X(i, j) - Y(i, j)]^2 \tag{4}$$

iii.   **Peak Signal to Noise Ratio (PSNR)**

The peak signal-to-noise ratio is always the inverse of the Mean Square Error (MSE). PSNR is a common metric for evaluating cypher image quality. More MSE and less PSNR are required for image security. PSNR can be calculated as follows [27].

$$\text{PSNR} = 10\log_{10} \frac{255}{\text{MSE}} \tag{5}$$

iv.   **UACI and NPCR**

Two tests are used to determine the sensitivity of the suggested encryption strategy in relation to the secret key and plain image: the number of pixels change rate (NPCR) and the unified average changing intensity (UACI) [28]. Equation 6 is the formula for calculating UACI.

$$\text{UACI} = \frac{1}{N} \left[\sum_{i,j} \frac{|C1(i, j) - C2(i, j)|}{255} \right] \tag{6}$$

where m denotes the number of rows, n the number of columns, and $C1(i, j)$ and $C2(i, j)$, respectively, the original and cypher image. Equation 7 can be used to calculate NPCR.

$$\text{NPCR} = \frac{\sum_{i,j} D(i, j)}{M \times N} \times 100\% \tag{7}$$

where m denotes the number of rows and n denotes the number of columns, then $D(i, j)$ is defined as follows:

$$D(i, j) = \begin{cases} 1, C1(i, j) \neq C2(i, j) \\ 0, otherwise \end{cases} \tag{8}$$

where $C1(i, j)$ and $C2(i, j)$ correspond to the original and cypher image, respectively.

e. Universal Image Quality Index (UIQ)

Universal index quality is used for calculating similarity between original image and cipher image. Range of UIQ is $[-1, 1]$ where value 1 indicates more similarity and value -1 indicates less similarity. UIQ is defined as in [28].

$$UQI(x, y) = \frac{\sigma xy}{\sigma x \sigma y} * \frac{2\mu x \mu y}{\mu x^2 + \mu y^2} * \frac{2\sigma x \sigma y}{\sigma x^2 + \sigma y^2} \tag{9}$$

where μx, μy, σx, σy and σxy are the mean of x & y, variance x & y and the covariance of x and y respectively.

f. Structural Similarity Index Measure (SSIM)

The SSIM is the extended version of the UIQ index. Range of SSIM is $[-1, 1]$ where value 1 indicates more similarity and value -1 indicates less similarity. SSIM is defined as in [28]

$$SSIM(x, y) = \left[\frac{(2\mu x \mu y + C1)(2\sigma xy + C2)}{(\mu x^2 + \mu y^2 + C1)(\sigma x^2 + \sigma y^2 + C2)} \right] \tag{10}$$

where $C1$, $C2$ are two constants and are used to stabilize the division with weak denominator.

g. Key space analysis

It is generally known that a good encryption scheme should have big key space and be sensitive to the key. In this algorithm, 256-bit key stream based on the plain image is designed, and all initial conditions, parameters are produced by the secret key. Confusion k1, Brownian motion R, diffusion k2 are all keys. Our proposed system key space is greater than 10^{128}. Therefore, a sufficiently large key space makes the brute force attack infeasible and is ensured for practical applications.

h. Histogram analysis

Image histogram graphically represents the distribution of pixels based on their occurrence. For highly secured encryption technique, histogram of cipher image should be uniformly distributed. Table 1 shows the histograms of the various medical original image and the corresponding cipher images. It can be observed that the histogram of the encrypted image is very close to uniform distribution.

Table 1 Histogram analysis of proposed system

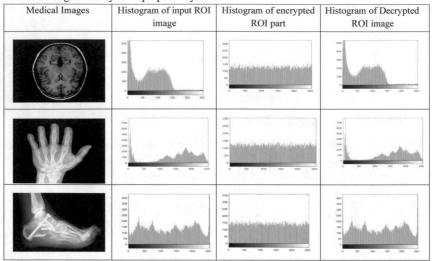

4.1   Results and Discussion

The experiments are run on a PC with an Intel I7-HQ 7700HQ processor, 16 GB of RAM, with MATLAB 2019 installed. In this experiment, we took 512 × 512 images of various sizes.

In our proposed scheme, only the ROI of the plain image is encrypted, and thus we examine only the histogram of the ROI. Table 1 illustrates the ROI histograms of the plain images, cipher images, and decrypted encrypted image are evenly distributed, similar to white noise. This shows that we have successfully changed the distribution relationship of pixel values, and our algorithm can effectively resist statistical attack.

Table 2 illustrates the original image, encrypted image and decrypted image. In the encrypted images clear shows that only region of interest part is encrypted that will reduce the complexity and less time takes to encrypt.

From Table 3 we concluded that the entropy values of cipher images are higher than the original pure image. MSE values are incremented depending on the image that indicates the extent of the encryption. Due to the proposed selective encryption method, the NPCR values are not varied very much, which suggests that the reduction in computational effort and time, as well as the similarity index also decrease to 0, which means that the value is higher, the dissimilarity between the input image and the encrypted ROI image (Fig. 6 and Table 4).

The encryption time has been greatly decreased utilising our technology, the authenticity of the image uploaded to the cloud has been verified, and security is provided using a two-level encryption scheme. We evaluated several metrics such as NPCR, MSE, PSNR, SSIM, Encryption time, and others to validate the effectiveness of our selective encryption system, and compared their values to those obtained using

Table 2 Input, ROI encrypted and decrypted images of proposed system

Medical Images	Histogram of input ROI image	Histogram of encrypted ROI part
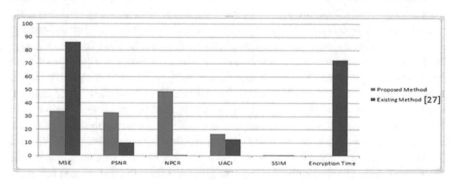		

Table 3 Performance parameters for proposed ROI encrypted system

Image Name	Entropy_In (Bit)	Entropy_Enc (Bit)	MSE	PSNR (db)	NPCR (%)	UACI (%)	UQI	SSIM
Leg	4.7639	4.89012	45.8391	31.9150	44.2445	12.4808	0.8668	0.5629
MRI	4.5597	4.99973	45.7087	22.8534	56.0204	26.6988	0.7826	0.3418
Hand	4.4401	5.02011	69.6292	29.8942	49.0779	16.1082	0.8421	0.5508

Fig. 6 Graphical analysis of parameters comparison with existing method [27]

Table 4 Efficiency of proposed ROI encrypted system

Image name	Encryption time (s)	Time (%) saving compared to full image encryption
MRI	0.217435	49.6094
Hand	0.315953	47.6563
Foot	0.298069	37.5000

Table 5 Parameters comparison with existing method for MRI Image

Parameters	Proposed method	Existing method [26]
MSE	45.7087	33.7087
PSNR	22.8534	32.8534
NPCR	56.0204	49.0204
UACI	26.6988	16.6988
SSIM	0.3418	0.4418
Encryption time	0.217435	0.317435

existing methods, as shown in Table 5. We discovered that our selective encryption strategy is a good method because it produces better outcomes than other methods.

5 Conclusion

This article introduces an automated area of interest (ROI) encryption method for applying security to medical images. The suggested technique is divided into three parts: an automatic ROI detection system, a permutation process, and a diffusion process. The position of each pixel in the ROI image that was changed using the Arnold map sequence and the pixel values are manipulated using the Brownian map. Our study's main contribution is the introduction of automatic ROI detection system that can helps to real time fast secure communication. For experimental validation of the proposed algorithm, a variety of metrics are used, including histogram analysis, Shannon entropy analysis, NPCR analysis, and UACI analysis, key space analysis and speed analysis. Finally, a comparison of our method to several well-known algorithms demonstrates its superiority. The results of the investigation show that the system can be used to encrypt ROI part of medical images effectively. In future research, In future work we want to research new encryption methods that can be effectively applied to the medical images in order to increase security and achieve a lower computing time, as desired by real-time applications.

References

1. Laiphrakpam, D.S., Khumanthem, M.S.: Medical image encryption based on improved elgamal encryption technique. Optik **147**, 88–102 (2017)
2. Kanso, A., Ghebleh, M.: An efficient and robust image encryption scheme for medical applications. Commun. Nonlinear Sci. Numer. Simul. **24**(1–3), 98–116 (2015)
3. Çavuşoğlu, Ü., Kaçar, S., Pehlivan, I., Zengin, A.: Secure image encryption algorithm design using a novel chaos based s-box. Chaos Solitons Fractals **95**, 92–101 (2017)
4. Artiles, J.A., Chaves, D.P., Pimentel, C.: Image encryption using block cipher and chaotic sequences. Signal Process Image Commun. **79**, 24–31 (2019)
5. Xingyuan, W., Junjian, Z., Guanghui, C.: An image encryption algorithm based on zigzag transform and LL compound chaotic system. Opt Laser Technol. **119**, 105581 (2019)
6. Liu, H., Wang, X.: Color image encryption based on one-time keys and robust chaotic maps. Comput. Math. Appl. **59**(10), 3320–3327 (2010)
7. Dong, C.: Color image encryption using one-time keys and coupled chaotic systems. Signal Process Image Commun. **29**(5), 628–640 (2014)
8. Khedr, W.I.: A new efficient and configurable image encryption structure for secure transmission. Multimed. Tools Appl. 1–25 (2019)
9. Boussif, M., Aloui, N., Cherif, A.: Smartphone application for medical images secured exchange based on encryption using the matrix product and the exclusive addition. IET Image Process **11**(11), 1020–1026 (2017)
10. Hua, Z., Yi, S., Zhou, Y.: Medical image encryption using high-speed scrambling and pixel adaptive diffusion. Signal Process **144**, 134–144 (2018)
11. Abd-El-Atty, B., Amin, M., Abd-El-Latif, A., Ugail, H., Mehmood, I.: An efficient cryptosystem based on the logistic-Chebyshev map. In: 2019 13th International Conference on Software, Knowledge, Information Management and Applications (SKIMA). IEEE (2019)
12. Zain, J.M., Clarke, M.: Reversible region of non-interest (RONI) watermarking for authentication of DICOM images. Int. J. Comput. Sci. Netw. Secur. **7**(9), 19–28 (2007)
13. Kundu, M.K., Das, S.: Lossless ROI medical image watermarking technique with enhanced security and high payload embedding. In: 2010 20th International Conference on Pattern Recognition (ICPR), pp. 1457–1460 (2010)
14. Al-Qershi, O.M., Bee, E.K.: ROI-based tamper detection and recovery for medical images using reversible watermarking technique. In: 2010 I.E. International Conference on Information Theory and Information Security (ICITIS), pp. 151–155 (2010)
15. Guo, X., Zhuang, T-G.: A region-based lossless watermarking scheme for enhancing security of medical data. J. Dig. Imaging **22**(1), 53–64 (2009)
16. Navas, K.A., Thampy, S.A., Sasikumar, M.: EPR hiding in medical images for telemedicine. Int. J. Biol. Life Sci. **3**(1), 44–47 (2007)
17. Fotopoulos, V., Stavrinou, M.L., Skodras, A.N.: Medical image authentication and self-correction through an adaptive reversible watermarking technique. In: 8th IEEE International Conference on BioInformatics and BioEngineering (BIBE), pp. 1–5 (2008)
18. Hyung-Kyo, L., Hee-Jung, K., Ki-Ryong, K., Jong-Keuk, L.: ROI medical image watermarking using DWT and Bit-plane. In: Asia-Pacific Conference on Communications, pp. 512–515, Perth, Western Australia (2005)
19. Gunjal, B.L., Mali, S.N.: ROI based embedded watermarking of medical images for secured communication in telemedicine. Int. J. Comput. Commun. Eng. 293–298 (2012)
20. Lin, C.-H., Yang, C.-Y., Chang, C.-W.: Authentication and Protection for Medical Image, vol. 6422, pp. 278–287. Springer, Berlin (2010)
21. Memon, N.A., Gilani, S.A.M.: NROI watermarking of medical images for content authentication. In Multi Topic Conference, 2008. INMIC 2008. IEEE International, pp. 106–110 (2008)
22. Amin, M., Abd El-Latif, A.A.: Efficient modified RC5 based on chaos adapted to image encryption. J. Electron. Imaging **19**(1), 013012 (2010)

23. Belazi, A., Abd El-Latif, A.A., Rhouma, R., Belghith, S.: Selective image encryption scheme based on DWT, AES S-box and chaotic permutation. In: 2015 International Wireless Communications and Mobile Computing Conference (IWCMC), pp. 606–610. IEEE (2015)
24. Tsafack, N., Sankar, S., Abd-El-Atty, B., Kengne, J., Jithin, K.C., Belazi, A., Mehmood, I., Bashir, A.K., Song, O.-Y., Abd El-Latif, A.A.: A new chaotic map with dynamic analysis and encryption application in internet of health things. IEEE Access 8, 137731–137744 (2020)
25. Abd EL-Latif, A.A., Abd-El-Atty, B., Abou-Nassar, E.M., Venegas-Andraca, S.E.: Controlled alternate quantum walks based privacy preserving healthcare images in internet of things. Opt. Laser Technol. 124, 105942 (2020)
26. Madhusudhan, K.N., Sakthivel, P.: A secure medical image transmission algorithm based on binary bits and Arnold map. J. Ambient Intell. Human. Comput. 12, 5413–5420 (2021). https://doi.org/10.1007/s12652-020-02028-5
27. Ahmad, J., Ahmed, F.: Efficiency analysis and security evaluation of image encryption schemes. Int. J. Video Image Process. Netw. Secur. 12, 18–31 (2012)
28. Wu, Y., Noonan, J.P., Agaian, S.: NPCR and UACI randomness tests for image encryption. Cyber J.: Multidiscip. J. Sci. Technol. J. Sel. Areas Telecommun. 31–38 (2011)

Multistability Analysis and MultiSim Simulation of A 12-Term Double-Scroll Hyperchaos System with Three Nonlinear Terms, Bursting Oscillations and Its Cryptographic Applications

Aceng Sambas, Sundarapandian Vaidyanathan, Sen Zhang,
Ahmed A. Abd El-Latif, Mohamad Afendee Mohamed,
and Bassem Abd-El-Atty

Abstract Using three nonlinear terms (two quadratics and one cubic), a new 12-term hyperchaos system with a double-scroll attractor is proposed in this chapter. Using bifurcation analysis, dynamical properties are exemplified and multistability property with coexisting hyperchaos attractors is established. MultiSim circuit simulation of the new hyperchaos system aids invalidation and practical application. Secure communication devices make use of dynamic systems having hyperchaos properties. Due to the good nonlinear properties of the presented hyperchaos system, we introduce its applications in designing secure substitution boxes (S-boxes) and

A. Sambas (✉)
Department of Mechanical Engineering, Universitas Muhammadiyah Tasikmalaya, Tasikmalaya,
Indonesia
e-mail: acengs@umtas.ac.id

S. Vaidyanathan
Research and Development Centre, Vel Tech University, Avadi, Chennai, India

S. Zhang
School of Physics and Opotoelectric Engineering, Xiangtan University, Hunan, Xiangtan 411105,
China

A. A. Abd El-Latif
EIAS Data Science Lab, College of Computer and Information Sciences, Prince Sultan
University, Riyadh 11586, Saudi Arabia
e-mail: aabdellatif@psu.edu.sa

Department of Mathematics and Computer Science, Faculty of Science, Menoufia University,
Shibin El Kom 32511, Egypt

M. A. Mohamed
Faculty of Informatics and Computing, Universiti Sultan Zainal Abidin, Terengganu, Malaysia

B. Abd-El-Atty
Department of Computer Science, Faculty of Computers and Information, Luxor University,
Luxor 85957, Egypt
e-mail: bassem.abdelatty@fci.luxor.edu.eg

generating pseudo-random number generators (PRNGs). Analyses of results prove the effectiveness of the presented S-box and PRNG mechanisms.

Keywords Double-scroll systems · Hyperchaotic systems · Hyperchaos · MultiSim design · Chaos-based S-box · Chaos-based PRNG · Chaos-based cryptography

1 Introduction

Engineering systems feature a good number of applications of dynamical systems having chaos or hyperchaos properties. The chaos applications arise in diverse domains such as microelectromechanical systems [1–3], wireless mobile robot [4], double/triple pendulum [5, 6], quarter-car vehicle model [7, 8], rail vehicle system [9], magnetorheological suspension system [10], circular mesh antenna [11], FitzHugh-Nagumo neurons models [12], Izhikevich neuron model [13], circuit models [14–16], Hindmarsh-Rose neuron model [17], steganography [18], Field-Programmable Gate Array (FPGA) [19], Moore-Spiegel thermo-mechanical [20], etc.

Dynamical systems having $p \geq 2$, where p designates the number of positive Lyapunov exponent are categorized as hyperchaotic systems [21], which exhibit characteristics suitable for engineering applications such that of secure strategies [22, 23], encryption [24, 25], memristive analog circuit [26–28], etc.

Several studies related to cybersecurity systems using chaotic systems in big data have developed significantly. In 2016, Sun proposed chaotic map model for telecare medical information systems. He has shown that the Chebyshev chaotic maps schema has a good performance in Telecare Medical Information Systems (TMIS) [29]. In 2020, Yim investigated a chaotic system synchronization for IoT Security Channel [30]. Also, in 2020, García-Guerrero et al. studied chaotic map for image encryption using wireless communication channel. They have used PIC-microcontroller via Zigbee channels in this study [31]. In the same year, Meshram et al. investigated fractional chaotic map for secure communication in Internet of Thing. They have designed new short signature scheme and show that the Short Signature Scheme (SSS) offers a better security assurance than currently established signature schemes [32]. Also, some application in big data can be seen in [33–38].

In this chapter, a construction of a new 4-D double-scroll hyperchaos behavior with two quadratic, one cubic and five balance points is proposed. We illustrate the dynamic behavior of the hyperchaos system via MATLAB signal plots, balance points, bifurcation analysis, multi-stability, coexisting hyperchaotic attractors, etc.

For practical implementation of the hyperchaos systems, designs carried out via electronic circuits [39, 40] or FPGA [41, 42] are immensely useful. In this chapter, we exhibit MultiSim circuit design of the proposed double scroll hyperchaos system with five balance points.

Designing secure S-box and PRNG algorithms show the significant performance in the development of modern cryptographic applications [43–46]. Chaos systems are considered a backbone in designing S-box and generating PRNG mechanisms [47–49]. Due to the good nonlinear properties of the presented hyperchaos system, we introduce its applications in designing secure S-boxes and generating PRNGs. Analyses of results prove the effectiveness of the presented S-box and PRNG mechanisms.

2 A Novel Double-Scroll Hyperchaos Dynamical System with Five Balance Points

A novel dynamical system is described in this work via the 4-D dynamics

$$\begin{cases} \dot{\xi}_1 = \alpha(\xi_2 - \xi_1) + \beta\xi_2\xi_3 + \xi_4 \\ \dot{\xi}_2 = \xi_1 + \gamma\xi_2 - \xi_1\xi_3^2 - \xi_4 \\ \dot{\xi}_3 = -\varepsilon\xi_3 + \xi_1\xi_2 \\ \dot{\xi}_4 = \delta(\xi_1 + \xi_2) \end{cases} \tag{1}$$

The four-dimensional vector $X = (\xi_1, \xi_2, \xi_3, \xi_4)$ terms the variable of the system (1), consisting of two quadratic nonlinearities and a cubic nonlinearity.

It will be verified by Lyapunov Exponents analysis in MATLAB platform. The system (1) shows the hyperchaos behavior when the parameters values

$$\alpha = 32, \ \beta = 18, \ \gamma = 15, \ \delta = 0.4, \ \varepsilon = 4 \tag{2}$$

For the time-series analysis of the all variable $X = (\xi_1, \xi_2, \xi_3, \xi_4)$ in system (1), we define the initial conditions as

$$\xi_1(0) = 0.2, \ \xi_2(0) = 0.3, \ \xi_3(0) = 0.3, \ \xi_4(0) = 0.2 \tag{3}$$

Using Wolf's procedure [50], the LE values of the 4-D hyperchaos behavior model (1) are mathematically computed for $T = 1e4$ seconds (see Fig. 1) as the followings: $\mu_1 = 3.4066$, $\mu_2 = 0.0413$, $\mu_3 = 0$ and $\mu_4 = -24.3438$. The existence of two positive LE values (viz. μ_1, μ_2) summarily signifies the hyperchaos nature of the model (1). That the sum of all LE values throughout the LE spectrum are observably to be negative, the model (1) is said to have dissipative motion in all of its trajectories which consequently converge to the hyperchaotic attractor.

Consider model (1) with the set of parameters and its values to be $(\alpha, \beta, \gamma, \delta, \varepsilon) = (32, 18, 15, 0.4, 4)$ and $X(0) = (0.2, 0.3, 0.3, 0.2)$, the LE values and the respective phase plot can be shown as in Figs. 1 and 2. A simple study of the signal plots in

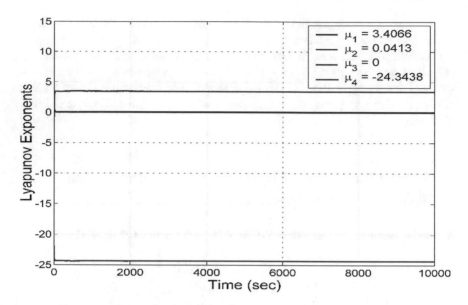

Fig. 1 Spectrum of LE for the double-scroll hyperchaos system (1)

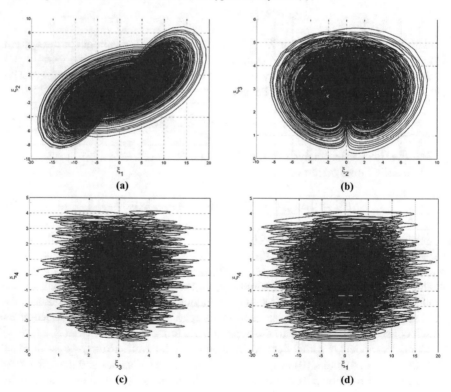

Fig. 2 MATLAB plots of the double-scroll hyperchaos system (1)

Fig. 2 reveals that the model (1) is equipped with a double-scroll hyperchaos attractor in the 4-D space.

The balance points of the 4-D hyperchaos model (1) for $(\alpha, \beta, \gamma, \delta, \varepsilon) =$ (32, 18, 15, 0.4, 4) are numerically evaluated in MATLAB as follows:

$$K_0 = \begin{bmatrix} 0 \\ 0 \\ 0 \\ 0 \end{bmatrix}, \ K_1 = \begin{bmatrix} 5.392 \\ -5.392 \\ -7.267 \\ -360.299 \end{bmatrix}, \ K_2 = \begin{bmatrix} -5.392 \\ 5.392 \\ -7.267 \\ 360.299 \end{bmatrix},$$

$$K_3 = \begin{bmatrix} 6.552 \\ -6.552 \\ -10.732 \\ -646.362 \end{bmatrix}, \ K_4 = \begin{bmatrix} -6.552 \\ 6.552 \\ -10.732 \\ 646.362 \end{bmatrix} \tag{4}$$

Using the Lyapunov stability theory [51], it can be found that the K_0 is a saddle point, while K_i, $(i = 1, 2, 3, 4)$ are saddle-foci points. This indicates that all the five balance points of the model (1) have instability nature and the double-scroll hyperchaos attractor is self-excited.

The rotational symmetry of the double-scroll hyperchaos system (1) is ξ_3—axis. Hence, there are twin trajectories for all non-zero initial states.

3 Multistability Property and Bifurcation Plots for the Double-Scroll Hyperchaos System

3.1 Bifurcation Analysis

Suppose that $\beta = 18$, $\gamma = 15$, $\delta = 0.4$, $\varepsilon = 4$, the initial state $X_0 = (0.2, 0.3, 0.3, 0.2)$, and alter α from 30 to 36. The bifurcation diagram of the state variable ξ_1 and the corresponding first three LE values are shown in Fig. 3a, b, respectively. Figure 3a signifies the fact that the double-scroll system (1) is hyperchaos except for several very narrow chaotic regions.

Next, we fix $\alpha = 30$, $\gamma = 15$, $\delta = 0.4$, $\varepsilon = 4$, the initial condition $X_0 = (0.2, 0.3, 0.3, 0.2)$, and alter β in the range of [14, 20]. Figures 4a, b show the bifurcation diagram of the state variable Z and the LE values spectrum. From Fig. 4b, it is obvious that the double-scroll system (1) exhibits hyperchaos with two positive LE values, chaos with one positive LE value and periodic orbit with zero largest LE value in the whole region. Moreover, it can be also noticed that there are many periodic windows as displayed in Fig. 4a.

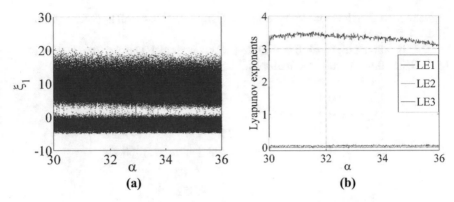

Fig. 3 Dynamical analysis of the double-scroll system (1) with respect to α: **a** the bifurcation diagram; **b** the corresponding LE values spectrum

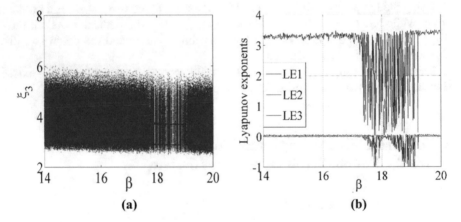

Fig. 4 Dynamical analysis of the double-scroll system (1) **a** the bifurcation diagram; **b** the corresponding LE values spectrum

3.2 Multistability

The term multistability is defined by a change attractors or stability system with having different initial conditions and commonly found in many nonlinear systems [52, 53]. Let the parameters be fixed as $\beta = 4$, $\gamma = 5$, $\delta = 5$ and $\varepsilon = 4$. We alter α in the interval [20, 24] and the initial states are picked as $X_0 = (0.2, 0.3, 0.3, 0.2)$ and $Y_0 = (-0.2, -0.3, 0.3, -0.2)$. The coexisting bifurcation diagrams of the state variable ξ_4 and the corresponding LE values with the initial state X_0 are plotted in Fig. 5a, b, respectively. As can be seen from Fig. 5, the double-scroll system (1) displays complex dynamics, such as chaos, hyperchaos, period, coexisting attractors as well as period-doubling bifurcation route to chaos. For example, $\alpha = 20$ and

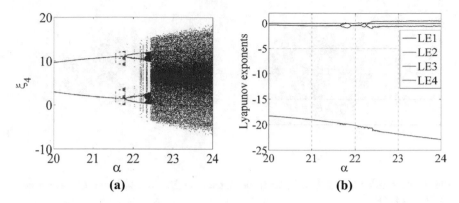

Fig. 5 Coexisting dynamics analysis of the double-scroll system with respect to α : **a** the coexisting bifurcation diagrams; **b** the corresponding LE values with $X_0 = (0.2, 0.3, 0.3, 0.2)$

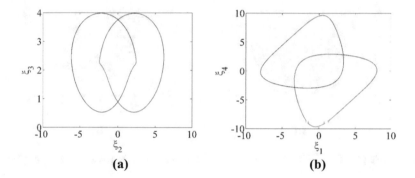

Fig. 6 Phase portraits of the coexisting period-1 attractors with $\alpha = 20$ in **a** the $\xi_2 - \xi_3$ plane; **b** the $\xi_1 - \xi_4$ plane

$\alpha = 22.5$, the system (1) shows two types of coexisting period-1 attractors and two kinds of chaotic attractors as shown in Figs. 6 and 7, respectively.

3.3 Bursting Oscillations

Multi bursting oscillations are used to describe a type of important electrical activity phenomenon used for information transmission and exchange in biological neurons and have been discovered in many nonlinear systems [54, 55]. Interestingly, when the parameters are taken as $\alpha = 32, \beta = 18, \gamma = 2.5, \delta = 0.4, \varepsilon = 4$, and the initial state is picked as $X_0 = (0.2, 0.3, 0.3, 0.2)$, the double-scroll system (1) exhibits chaotic bursting behaviors as shown in Fig. 8. Furthermore, when the parameters are taken as $\alpha = 16, \beta = 10, \gamma = 1, \delta = 0.4, \varepsilon = 4$, and the initial conditions are selected

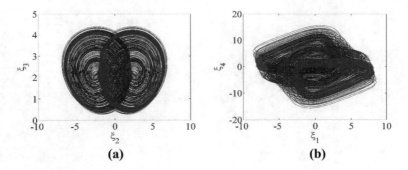

Fig. 7 Phase portraits of the coexisting chaotic attractors with $\alpha = 22.5$ in **a** the $\xi_2 - \xi_3$ plane; **b** the $\xi_1 - \xi_4$ plane

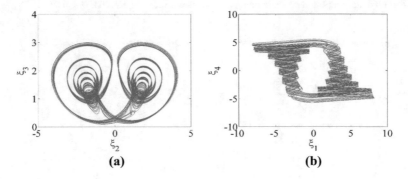

Fig. 8 Phase portraits of the chaotic bursting attractor in **a** the $\xi_2-\xi_3$ plane; **b** the $\xi_1-\xi_4$ plane

as $X_0 = (0.2, 0.3, 0.3, 0.2)$ and $Y_0 = (-0.2, -0.3, 0.3, -0.2)$, respectively, the double-scroll system (1) can depict a striking phenomenon of coexisting periodic bursting behaviors, which has not been observed in other similar hyperchaos systems. The typical bursting phase portraits are displayed in Fig. 9.

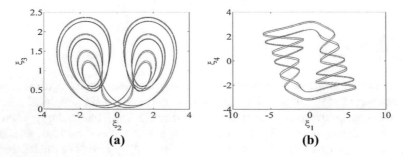

Fig. 9 Phase portraits of the coexisting periodic bursting attractors in **a** the $\xi_2-\xi_3$ plane; **b** the $\xi_1-\xi_4$ plane

4 Electronic Circuit by MultiSIM Software

This subchapter is meant for the construction of an electronic circuit for the new double-scroll hyperchaos system as described in (1). Figure 10 is the representation of system (1) having four channels electronic circuit with variables $\xi_1, \xi_2, \xi_3, \xi_4$ and that its dynamics can be described by (5).

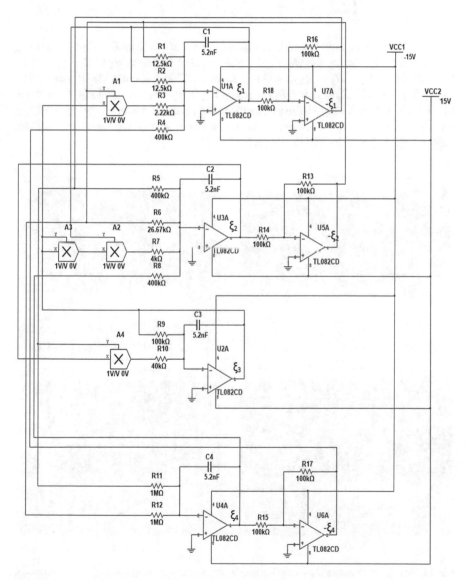

Fig. 10 Circuit design for the new hyperchaotic two-scroll system

$$\begin{cases} C_1\dot{\xi}_1 = \dfrac{1}{R_1}\xi_2 - \dfrac{1}{R_2}\xi_1 + \dfrac{1}{10R_3}\xi_2\xi_3 + \dfrac{1}{R_4}\xi_4 \\[2mm] C_2\dot{\xi}_2 = \dfrac{1}{R_5}\xi_1 + \dfrac{1}{R_6}\xi_2 - \dfrac{1}{100R_7}\xi_1\xi_3^2 - \dfrac{1}{R_8}\xi_4 \\[2mm] C_3\dot{\xi}_3 = -\dfrac{1}{R_9}\xi_3 + \dfrac{1}{10R_{10}}\xi_1\xi_2 \\[2mm] C_4\dot{\xi}_4 = \dfrac{1}{R_{11}}\xi_1 + \dfrac{1}{R_{12}}\xi_2 \end{cases} \tag{5}$$

Here, $\xi_1, \xi_2, \xi_3, \xi_4$ are the output voltages of the capacitors C_1, C_2, C_3 and C_4, respectively. We choose the values of the circuital elements as $R_1 = R_2 = 12.5$ kΩ, $R_3 = 2.22$ kΩ, $R_4 = R_5 = R_8 = 400$ kΩ, $R_6 = 26.67$ kΩ, $R_7 = 4$ kΩ, $R_{10} = 40$ kΩ, $R_{11} = R_{12} = 1$ MΩ, $R_9 = R_{13} = R_{14} = R_{15} = R_{16} = R_{17} = R_{18} = 100$ kΩ, $C_1 = C_2 = C_3 = C_4 = 5.2$ nF. Using Multisim platform, the phase portraits on the oscilloscope are shown in Fig. 11. We observe a very good consistency between the Multisim output (Fig. 11) and the MATLAB plots (Fig. 2).

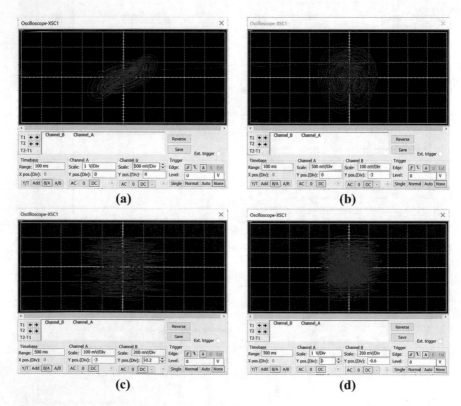

Fig. 11 MultiSIM chaotic attractors of the new hyperchaotic two-scroll system **a** $x_1 - x_2$ plane, **b** $x_2 - x_3$ plane, **c** $x_3 - x_4$ plane and **d** $x_1 - x_4$ plane

5 PRNG Algorithm

Generating PRNGs play a critical task in designing modern stream ciphers. Based on the effectiveness of the presented hyperchaos system, we presented a new PRNG algorithm. The detailed steps of the proposed PRNG algorithm are outlined below.

Step 1: Select the initial state $X_0 = (\zeta_1, \zeta_2, \zeta_3, \zeta_4)$ and control parameters (β, γ, δ, and ε) for solving the hyperchaos system (1) to produce four sequences ($S1$, $S2$, $S3$, and $S4$).

Step 2: Transform the elements of $S1$ into integers in the range [0, 2] as R.

$$R = \mathrm{round}\left(S1 \times 10^4 \bmod 3\right) \tag{6}$$

Step 3: Construct sequence Q from the last three sequences utilizing R sequence.

$$Q_i = \begin{cases} S2_i \text{ if } R_i = 0 \\ S3_i \text{ if } R_i = 1 \\ S4_i \text{ if } R_i = 2 \end{cases} \tag{7}$$

Step 4: Transform the elements of Q into integers in the range [0, 255] as PRNG sequence.

$$\mathrm{PRNG} = \mathrm{round}\left(Q \times 10^{12} \bmod 256\right) \tag{8}$$

To validate the randomness property of the generated PRNG sequence, we perform NIST SP 800-22 tests. The fundamental role of these tests is to detect any non-randomness property existing in a sequence. The P-value of each test is in period [0, 1]. If the P-value is greater than the threshold $\alpha = 0.01$, this means that the sequence passes this test [56]. The initial states and control parameters are set as: $X_0 = (0.2, 0.305, 0.379, 0.2)$, $\beta = 18$, $\gamma = 15$, $\delta = 0.4$, and $\varepsilon = 4$. The results of NIST SP 800-22 tests are given in Table 1, in which the generated sequence passed all NIST SP 800-22 tests for measuring randomness.

6 S-Box Scheme

The designing of S-boxes for reliable cryptographic ciphers attracts a great deal of attention from most cryptographic researchers. Admittedly, S-boxes are an important component of most modern block ciphers. Based on the effectiveness of the presented hyperchaos system, we presented a new chaos-based S-box scheme. The detailed steps of the proposed S-box scheme are outlined below.

Step 1: As stated in PRNG algorithm Step 1.
Step 2: As stated in PRNG algorithm Step 2.

Table 1 Results of NIST SP
800-22 tests for the generated
PRNG sequence

Test		P-value	Passed
Long runs of ones		0.613273	✓
Linear complexity		0.556022	✓
Random excursions variant ($x = 1$)		0.732373	✓
Frequency		0.059836	✓
Random excursions ($x = 1$)		0.734235	✓
No overlapping templates		0.882052	✓
Overlapping templates		0.622666	✓
Block-frequency		0.513532	✓
Approximate entropy		0.078807	✓
Rank		0.931527	✓
Universal statistical		0.768934	✓
DFT		0.147084	✓
Runs		0.857512	✓
Cumulative sums	Reverse	0.029051	✓
	Forward	0.079951	✓
Serial	Test1	0.871239	✓
	Test2	0.869036	✓

Step 3:   As stated in PRNG algorithm Step 3.

Step 4:   Transform the elements of Q into integers in the range $[0, 2^N - 1]$.

$$T = \text{round}\left(Q \times 10^{12} \bmod 2^N\right) \tag{9}$$

Step 5:   Collect the first 2^N distinct element from sequence T as the constructed $N \times N$ S-box.

The initial states and control parameters that utilized to construct an 8×8 S-box are set as: $X_0 = (0.2, 0.305, 0.379, 0.2)$, $\beta = 18$, $\gamma = 15$, $\delta = 0.4$, and $\varepsilon = 4$. The generated S-box using the primary key parameters is provided in Table 2, while Table 3 presents a simple comparison of the performance of the presented S-box alongside related S-box schemes in terms of strict avalanche criterion (SAC), nonlinearity, bit independence (BIC), differential probability (DP), and linear approximation probability (LP), in which the proposed S-box scheme has good nonlinearity, SAC, and BIC properties.

Table 2 An 8 × 8 S-box constructed using the suggested mechanism

0	243	86	152	124	253	249	135	2	172	132	177	118	185	82	139
76	11	127	191	64	255	220	112	83	232	69	153	106	236	178	197
72	44	113	138	180	144	157	100	192	43	66	81	164	229	217	67
78	97	149	55	42	239	162	39	37	53	231	226	101	188	89	204
114	245	14	126	155	16	123	117	230	122	147	134	80	224	13	198
59	120	49	171	88	17	181	32	8	33	57	9	110	200	161	50
6	22	216	168	129	246	244	141	206	73	47	23	130	131	250	166
19	209	20	137	54	218	115	151	15	160	170	186	154	60	241	34
4	103	38	190	222	5	158	235	228	61	248	77	163	165	75	225
247	46	143	233	84	30	202	94	128	221	237	91	74	212	7	79
68	182	174	90	187	105	27	41	116	28	65	3	25	40	109	71
107	36	95	195	215	199	35	207	125	142	210	121	98	51	119	140
29	58	99	1	70	136	145	251	223	85	242	213	10	205	96	179
52	227	18	87	189	194	31	203	133	111	167	48	159	173	92	104
156	234	176	193	196	146	150	63	208	252	254	240	201	12	219	56
184	21	45	238	102	108	93	214	211	183	169	148	62	175	26	24

Table 3 Simple comparison of the performance of the presented S-box alongside related S-box schemes

S-box scheme	Nonlinearity	SAC	BIC-SAC	BIC-NL	DP	LP
Proposed	106.25	0.4988	0.4988	103.9	0.0469	0.1172
Abd EL-Latif et al. [43]	106.25	0.5037	0.5065	103.7	0.0391	0.1016
Abd El-Latif et al. [45]	106	0.4958	0.5023	103.9	0.0313	0.1250
Abd-El-Atty et al. [48]	105.5	0.5032	0.5051	104.0	0.03906	0.1172
Abd EL-Latif et al. [49]	106	0.4993	0.5030	104.2	0.0391	0.1250
Khan and Asghar Belazi et al. [57]	102.00	0.5178	0.4999	102.9	0.0313	0.1250
Belazi et al. [58]	105.25	0.4956	0.4996	103.8	0.0391	0.1562
Belazi and Abd El-Latif [59]	105.50	0.5000	0.4970	103.8	0.0468	0.1250

7 Conclusions

In this chapter, making use of three nonlinear terms (two quadratics and one cubic), a new twelve-term hyperchaos system with a double-scroll attractor was proposed and illustrated with MATLAB signal plots. Using bifurcation analysis and multi-stability analysis, various dynamical properties were explored, and it was shown that the double-scroll hyperchaos system has a self-excited attractor and five unstable balance points. Also, system exhibits chaotic bursting behaviors and coexisting periodic bursting behaviors. MultiSim circuit simulation of the new hyperchaos system was designed for validation of the new hyperchaos system and practical applications. Additionally, we introduced a secure S-box mechanism and a robust PRNG algorithm based on the presented hyperchaos system, in which analyses of results proved the effectiveness of the presented S-box and PRNG mechanisms and the reliability of the presented hyperchaos model in designing different cryptographic applications.

Acknowledgements This work was supported by Luxor University and Menoufia University, Egypt.

References

1. Defoort, M., Rufer, L., Fesquet, L., Basrour, S.: Microsyst. Nanoeng. **7**, 1–11 (2021)
2. Luo, S., Li, S., Tajaddodianfar, F., Hu, J.: IEEE Sens. J. **18**, 3524–3532 (2018)
3. Liqin, L., Gang, T.Y., Zhiqiang, W.: J. Vib. Control **12**, 57–65 (2006)
4. Vaidyanathan, S., Sambas, A., Mamat, M., Sanjaya, M.S.M.: Arch. Control Sci. **27** 541–554 (2017)
5. Kumar, R., Gupta, S., Ali, S.F.: Mech. Syst. Signal Process. **124**, 49–64 (2019)
6. Jahn, B., Watermann, L, Reger, J.: Automatica **125**, 109403 (2021)
7. Chang, S.C., Hu, J.F.: Adv. Mech. Eng. **10**, 1687814018771764 (2018)
8. Naik, R.D., Singru, P.M.: Commun. Nonlinear Sci. Numer. Simul. **16**, 3397–3410 (2011)
9. Kaas-Petersen, C., True, H.: Veh. Syst. Dyn. **15**, 208–221 (1986)
10. Zhang, C., Xiao, J.: J. Comput. Nonlinear Dyn. **13**, 021007 (2018)
11. Wu, Q.L., Zhang, W., Dowell, E.H.: Int. J. Non-Linear Mech. **102**, 25–40 (2018)
12. Binczak, S., Jacquir, S., Bilbault, J.M., Kazantsev, V.B., Nekorkin, V.I.: Neural Netw. **19**, 684–693 (2006)
13. Nobukawa, S., Nishimura, H., Yamanishi, T., Liu, J.Q.: PloS one **10**, e0138919 (2015)
14. Sambas, A., Vaidyanathan, S., Tlelo-Cuautle, E., Abd-El-Atty, B., El-Latif, A.A.A., Guillen-Fernandez, O., Sukono, Hidayat, Y., Gundara, G.: IEEE Access **8**, 137116–137132 (2020)
15. Sambas, A., Vaidyanathan, S., Bonny, T., Zhang, S., Hidayat, Y., Gundara, G., Mamat, M.: Appl. Sci. **11**, 788 (2021)
16. Sambas, A., Vaidyanathan, S., Zhang, S., Zeng, Y., Mohamed, M.A., Mamat, M.: IEEE Access **7**, 115454–115462 (2019)
17. Wang, S., He, S., Rajagopal, K., Karthikeyan, A., Sun, K.: Eur. Phys. J. Spec. Top. **229**, 929–942 (2020)
18. Vaidyanathan, S., Sambas, A., Abd-El-Atty, B., Abd El-Latif, A.A., Tlelo-Cuautle, E., Guillén-Fernández, O., Mamat, M., Mohamed, M.A, Alcin, M., Tuna, M., Pehlivan, I., Koyuncu, I., Ibrahim, M.A.H.: IEEE Access **9**, 81352–81369 (2021)
19. Sambas, A., Vaidyanathan, S., Tlelo-Cuautle, E., Zhang, S., Guillen-Fernandez, O., Hidayat, Y., Gundara, G.: Electronics **8**, 1211 (2019)
20. Vaidyanathan, S., Sambas, A., Azar, A.T., Serrano, F.E., Fekik, A.: Backstepping Control of Nonlinear Dynamical Systems, pp. 165–189 (2021)
21. Vaidyanathan, S., Dolvis, L.G., Jacques, K., Lien, C.H., Sambas, A.: Int. J. Modell. Identif. Control **32**, 30–45 (2019)
22. Zhong, C., Pan, M.S.: Chin. J. Liquid Cryst. Disp. **35**, 91–97 (2020)
23. Li, Z., Jiang, A., Mu, Y.: Lect. Notes Electr. Eng. **571**, 796–810 (2020)
24. Ding, L., Ding, Q.: Electronics **9**, 1–20 (2020)
25. Shakiba, A.: Multimed. Tools Appl. **79**, 32575–32605 (2020)
26. Zhou, L., Wang, C., Zhou, L.: Nonlinear Dyn. **85**, 2653–2663 (2016)
27. Wang, X., Gao, M., Min, X., Lin, Z., Iu, H.H.C.: IEEE Access **8**, 182240–182248 (2020)

28. Raj, B., Vaidyanathan, S.: Stud. Comput. Intell. **701**, 449–476 (2017)
29. Sun, Y.: J. Inf. Hiding Multimed. Signal Process. **7**, 1006–1019
30. Yi, G.S.: J. Korea Inst. Electron. Commun. Sci. **15**, 981–986 (2020)
31. García-Guerrero, E.E., Inzunza-González, E., López-Bonilla, O.R., Cárdenas-Valdez, J.R., Tlelo-Cuautle, E.: Chaos Solitons Fractals **133**, 109646 (2020)
32. Meshram, C., Ibrahim, R.W., Obaid, A.J., Meshram, S.G., Meshram, A., Abd El-Latif, A.M.: J. Adv. Res. **32**, 139–148 (2020)
33. Abd El-Latif, A.A., Abd-El-Atty, B., Hossain, M.S., Elmougy, S., Ghoneim, A.: IEEE Access **6**, 10332–10340 (2018)
34. Gad, R., Talha, M., Abd El-Latif, A.A., Zorkany, M., Ayman, E.S., Nawal, E.F., Muhammad, G.: Fut. Gener. Comput. Syst. **89**, 178–191 (2018)
35. Zhang, W.Z., Elgendy, I.A., Hammad, M., Iliyasu, A.M., Du, X., Guizani, M., Abd El-Latif, A.A.: IEEE Internet Things J. **8**, 8119–8132 (2020)
36. Abd EL-Latif, A.A., Abd-El-Atty, B., Abou-Nassar, E.M., Venegas-Andraca, S.E.: Optics Laser Technol. **124**, 105942 (2020)
37. Abou-Nassar, E.M., Iliyasu, A.M., El-Kafrawy, P.M., Song, O.Y., Bashir, A.K., Abd El-Latif, A.A.: IEEE Access **8**, 111223–111238 (2020)
38. Abd EL-Latif, A.A., Abd-El-Atty, B., Venegas-Andraca, S.E., Mazurczyk, W.: Fut. Gener. Comput. Syst. **100**, 893–906 (2019)
39. Yang, L., Yang, Q., Chen, G.: Commun. Nonlinear Sci. Numer. Simul. **90**, 105362 (2020)
40. Zhou, W., Wang, G., Lu, H.H.C., Shen, Y., Liang, Y.: Nonlinear Dyn. **100**, 3937–3957 (2020)
41. Vaidyanathan, S., Tlelo-Cuautle, E., Anand, P.G., Sambas, A., Guillén-Fernández, O., Zhang, S.: Int. J. Autom. Control **15**, 128–148 (2021)
42. Vaidyanathan, S., Tlelo-Cuautle, E., Sambas, A., Dolvis, L.G., Guillén-Fernández, O.: Int. J. Comput. Appl. Technol. **64**, 223–234 (2020)
43. Abd EL-Latif, A.A., Abd-El-Atty, B., Venegas-Andraca, S.E.: Opt. Laser Technol. **116**, 92–102 (2019)
44. Abd El-Latif, A.A., Abd-El-Atty, B., Mazurczyk, W., Fung, C., Venegas-Andraca, S.E.: IEEE Trans. Netw. Serv. Manag. **17**(1), 118–131 (2020)
45. Abd El-Latif, A.A., Abd-El-Atty, B., Amin, M., Iliyasu, A.M.: Sci. Rep. **10**(1), 1–16 (2020)
46. Abd EL-Latif, A.A., Abd-El-Atty, B., Venegas-Andraca, S.E.: Phys. A: Stat. Mechan. Appl. **547**, 123869 (2020)
47. Tsafack, N., Kengne, J., Abd-El-Atty, B., Iliyasu, A.M., Hirota, K., Abd El-Latif, A.A.: Inf. Sci. **515**, 191–217 (2020)
48. Abd-El-Atty, B., Amin, M., Abd-El-Latif, A., Ugail, H., Mehmood, I.: 2019 13th International Conference on Software, Knowledge, Information Management and Applications (SKIMA), pp. 1–6. IEEE (2019)
49. Abd EL-Latif, A.A., Abd-El-Atty, B., Belazi, A., Iliyasu, A.M.: Electronics **10**(12), 1392 (2021)
50. Wolf, A., Swift, J.B., Swinney, H.L., Vastano, J.A.: Physica D **16**, 285–317 (1985)
51. Khalil, H.K.: Nonlinear Systems. Pearson, New York (2001)
52. Zhang, S., Zheng, J., Wang, X., Zeng, Z., He, S.: Nonlinear Dyn. **102**, 2821–2841 (2020)
53. Zhang, S., Zeng, Y., Li, Z., Wang, M., Xiong, L.: Chaos **28**, 013113 (2018)
54. Zhang, S., Wang, X., Zeng, Z.: Chaos **30**, 053129 (2020)
55. Zhang, S., Zeng, Y.: Chaos Solitons Fractals **120**, 25–40 (2019)
56. Alanezi, A., Abd-El-Atty, B., Kolivand, H., El-Latif, A., Ahmed, A., El-Rahiem, B., Sankar, S., Khalifa, H.: Secur. Commun. Netw. **2021** (2021)
57. Khan, M., Asghar, Z.: Neural Comput. Appl. **29**(4), 993–999 (2018)
58. Belazi, A., Khan, M., Abd El-Latif, A.A., Belghith, S.: Nonlinear Dyn. **87**(1), 337–361 (2017)
59. Belazi, A., Abd El-Latif, A.A.: Optik **130**, 1438–1444 (2017)

Printed in the United States
by Baker & Taylor Publisher Services